D1447853

Electroplating

A Treatise for the Beginner and for the
Most Experienced Electroplater

BY

HENRY C. REETZ

POPULAR MECHANICS
TWENTY-FIVE-CENT HANDBOOK SERIES

CHICAGO

POPULAR MECHANICS COMPANY

PUBLISHERS

Electroplating

by Henry C. Reetz

Original copyright in 1911 by
H. H. Windsor

Originally published in 1911 by
Popular Mechanics Co., Chicago

Reprinted by
Lindsay Publications Inc
Bradley IL 60915

ISBN 1-55918-008-0

1989

5   6   7   8   9   10

THIS book is one of the series of handbooks on industrial subjects being published by the Popular Mechanics Company. Like the Magazine, these books are "written so you can understand it," and are intended to furnish information on mechanical subjects at a price within the reach of all.

The texts and illustrations have been prepared expressly for this Handbook Series, by experts; are up-to-date, and have been revised by the editor of Popular Mechanics.

WARNING

Remember that the materials and methods described here are from another era. Workers were less safety conscious then, and some methods may be downright dangerous. Be careful! Use good solid judgement in your work, and think ahead. Lindsay Publications Inc. has not tested these methods and materials and does not endorse them. Our job is merely to pass along to you information from another era. Safety is your responsibility.

Write for a complete catalog of unusual books available from:

Lindsay Publications Inc
PO Box 12
Bradley IL 60915-0012

CONTENTS

———

ELECTROPLATING

CHAPTER I

INTRODUCTION

ELECTROPLATING, as commonly understood, is the coating of an inferior with a more valuable metal by means of electricity. It is scientifically defined as "the art or process of covering any electrically conducting material with an adherent and lasting film of metal, in a bath containing a solution of that metal, by means of the electrolytic action induced by a current from a battery or dynamo." This technical definition may seem rather difficult for the average reader to understand, but will be made plain as we proceed step by step with the subject.

It is not necessary that the electroplater, in making a beginning, should have a scientific knowledge of chemistry and electricity, although, unquestionably, the better posted he is in those branches of the sciences that relate to his trade, the more likely he is to succeed. While it is quite possible for a man of average skill to fit himself to be an expert electroplater in certain lines, by a careful study of methods and apparatus, and especially by imitating a practical plater, yet it goes without saying that, the more he can absorb of the real science of the business, the *why* of the *how*, the less will be his trouble with batteries and solutions and the greater will be his success.

In this handbook, therefore, we will endeavor to give brief and practical directions calculated to be of benefit to those already engaged, as well as to those about to engage, in the electroplating business, with no more technical detail than is necessary for practical work, and yet with such explicit directions concerning the actual operations as an old hand at the business thinks may be useful to the beginner.

Two things will be urged at the start, and, but for the necessity of being brief, they would be repeated on every page, for without them success cannot be expected. They are:

Use care at every step. Nowhere is carelessness more costly than in the electroplating shop. Constant vigilance is the price of success.

Study the why of things. Know why you do this and that, and you will be more apt to do it right. If you know what you are about at every stage of the process and do things right, the work will come out right, and all will be well. If you blunder along, hit or miss, you will be inquiring the price of junk before you are six months older.

OLD AND NEW METHODS

In electroplating, the apparatus which supplies the electric current, through the aid of which the film of metal is deposited on the article to be plated, is, of course, the essential feature. The ancients used to hammer out a thin leaf of gold or silver and solder it to a surface, often doing really excellent work, but this was a most tedious and expensive process. A kind of plating is also done by immersion in melted metal, usually a soft composition. This is not permanent and could not be employed with precious metals on account of the expense.

Modern plating, or real plating, is electroplating, by which we use a small amount of metal, deposit it evenly, thoroughly, and in a short time, and yet obtain lasting and most satisfactory results. Electric current does the work.

HOW METAL IS DEPOSITED

In simple forms of electroplating apparatus, the bath containing the metallic solution may itself form the battery, and copper is easily deposited in this manner. The

more common and practical plan, however, is to obtain
current from a source outside the bath, as from a regu-
larly constructed electric battery or dynamo.

When articles are to be electroplated, they are suspended
by wires in the plating tank, which contains a solution of
the metal which is to be deposited on them.   From another

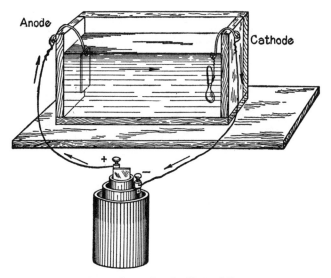

Fig. 1. Connections for Electroplating

wire is hung a piece of the same metal as that in the
solution.   The articles to be plated are connected by a
wire to the negative pole of a battery and the metal plate
is connected in the same way to the positive pole of the
battery.

Now the metallic solution in the tank will conduct the
electricity, and so a complete electric circuit is formed

by which the current flows from the positive pole of the battery through the connecting wire to the metal plate hanging in the tank. From the metal plate, it flows through the solution to the article to be plated and thence back through the other wire to the negative pole of the battery —completing the circuit. Fig. 1 shows the path of the current in an electroplating circuit.

The metal plate attached to the positive pole of the battery is called the "anode" and the objects to be plated in the tank are called "cathodes." The use of these terms can be understood if we think of the battery as being below

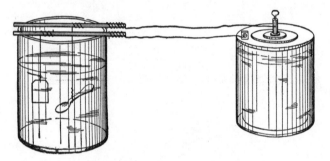

Fig. 2. Simple Electroplating Outfit

the plating tank, as shown in the illustration. "Anode" is a Greek word meaning "the way up"—from the battery to the solution—and "cathode" means "the way down." By the action of the current, particles of metal are attracted from the solution to the articles to be plated, and at the same time an equal amount of the same metal is thrown off the anodes into the solution. Thus, although the solution is being constantly robbed of metal, it is being fed at the same time, and so its strength is maintained. Of course, the anodes must be plates of the same metal as that contained in the solution; thus, silver anodes in a silver solution, nickel anodes in a nickel solution, etc. The process by which the electric current affects the decomposition of the metal to be deposited on the articles to be plated is called "electrolysis."

To make all this very plain, and to start the beginner at once with a practical working experiment, the diagram, Fig. 2, shows a very simple plating outfit which anyone can make at home. The explanations given will help the beginner to master the first principles of electroplating. A silver-plating outfit has been selected for illustration because silver is the metal most easily deposited and therefore is a favorite with experimenters.

HOW TO MAKE A SMALL SILVERPLATING OUTFIT.

For a tank take an ordinary glass fruit jar or any other receptacle of glass, not metal, which will hold 1 qt. of liquid. Fill it nearly full with rain or distilled water and then add ¾ oz. of silver chloride and 1½ oz. of chemically pure potassium cyanide. Let this dissolve and incorporate well with the water before using. This is the bath.

Take an ordinary wet battery and fasten two copper wires to the terminals, as in Fig. 1. Fasten the other ends of the wires to two rods of heavy copper wire or ¼-in. brass pipe, which are to be laid across the top of the "tank." The wires must be well soldered to the rods to make a good connection.

When the solution is ready, which is when the crystals are entirely dissolved, the outfit is ready for work. Procure a small piece of silver, a silver button, ring, chain, or anything made entirely of sterling silver, fasten a small copper wire to this, and hang it by this wire to the rod connected with the carbon of the battery. This forms the anode. The article to be plated is to be suspended in a similar manner from the other rod and will form the cathode.

Be sure that the article to be plated is chemically clean.

It may be cleaned by scrubbing with pumice and a brush saturated in water. When cleaning an article, there should always be a copper wire attached to it. It should never be touched by the hand after you have once started to clean it, because the touch of the fingers will deposit enough grease to cause the silver plate to peel off when finished. When well scoured, run clear, cold water over the article and, if it appears at all greasy, place it in hot water. When well cleaned, place it in the plating bath and carefully watch the results.

If small bubbles appear on the surface of the bath, you will know that you have too much of the anode immersed and you must draw out the piece until you can see no more bubbles. The surface of the anode should be about the same as the surface of the article to be plated. If the anode is too large, bubbles will appear, as stated; if too small, the metal contained in the solution will be exhausted.

Leave the piece to be plated in the solution for about half an hour, then take it out, clean off the yellowish scum with a toothbrush and some pumice, rinse in clean water and dry in sawdust. When it is thoroughly dry, take a cotton flannel rag and some polishing powder and polish the article. It must have a fine polish before plating if it is desired to have a finely polished surface after the plate is put on.

To see if your battery is working, take a small copper wire and touch one end to the anode rod and the other to the cathode rod. When these two ends are touched and separated, there should be a small spark.

This description applies only to silverplating. Articles of lead, pewter, tin or any other soft metal cannot be silverplated unless they have first been copperplated. This, as stated, is merely an experiment to interest the beginner and to form a sort of object lesson which will enable him to grasp more easily the details which follow and the reasons why.

CHAPTER II

ELECTRIC current used for electroplating is derived either from voltaic cells or from dynamo-electric generators, usually called "dynamos." If power is available, a dynamo is of course in the long run the most economical means of producing the electric current required for electroplating. But if a dynamo is out of the question, either for want of power or on account of the initial expense, its place can be very satisfactorily filled by batteries coupled together in sufficient number to produce the required current.

ELECTRIC BATTERIES

An electric battery is composed of several cells, although a cell is sometimes called a battery. The cells are of various kinds, wet and dry, and the wet cells may be of single or double liquid. The types described below refer to kinds or patterns of cells, and not to any special patents. There are as many special kinds of electric cells, or batteries, as there are manufacturing concerns, and even more. As a general rule, the higher-priced cells are likely to be worth all the money asked, and the cheaper ones really cost more in the end. This rule will apply pretty generally to chemicals and to all supplies. The beginner is advised to get as few articles as possible, and get the best; then to use them carefully, so as to get the greatest value out of them.

With regard to the different kinds of cells, the Smee and Wollaston are perhaps the most common types of single-liquid cells, and the Daniell, Fuller, Bunsen, and gravity cells may be taken as representative patterns of double-liquid cells. The principle consists in the exciting of an electric current by the solution (by an acid or an acid salt) of the readily attacked (or "negative") metal, in conjunction with a metal or other element (as carbon) not so readily attacked (this being called the "positive" metal or element), the effect being to produce what is called a "difference of potential" which sets up an "electromotive force" that tends to drive a current of electricity around the circuit when a passage is completed for it outside the cell.

The electric current has several properties, among which we may consider force and quantity, both of which are governed or influenced by the resistance of the bodies through which the current passes. This has often been compared to water in a pipe, which may be under a varying

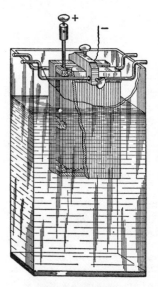

Fig. 3.   Smee Cell

pressure according to the level, and may have a flow according to the opening at the faucet. The difference of potential may be likened to the water level, the electromotive force to the pressure, and the current to the flow or volume of water. For some purposes, a very high voltage* is required, and for others, as in electroplating, a low voltage and a large volume of current are needed. We will now consider the construction of the cells, and then the methods of connecting them. to get the desired results.

THE SMEE CELL

The smee cell, Fig. 3, is a square glass jar having two zinc plates held together by a clamp, and between them, well insulated, a platinum plate, or a silver plate covered with "plat-

*The "volt" is the practical unit of electromotive force. The "ohm" is the unit of resistance. The "ampere" is the unit of current.

inum black," which is a preparation of finely divided platinum, black in color, obtained by the action of potassium hydrate on platinum chloride.

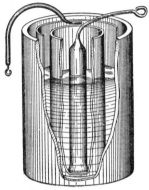

Platinum black has the property of absorbing or occluding gases, and is employed in the Smee cell for this purpose. The plates must have sufficient room so as not to come in contact with the zinc sulphate that forms by the dissolving of the plates. The vessel is filled with a solution of sulphuric acid (by measure, 1 part of the acid to 7 parts of water) and is then ready for use. Connections are made by a small copper wire, No. 16 gauge, with the zinc element to the cathode rod, and from the plat-

Fig. 4. Daniell Cell

inum element to the anode rod. Several cells may be placed together in a frame and connected to form a battery. The Smee cell has the advantage of being able to give a fairly steady current for a long time, as the platinum black prevents the accumulation on the positive element of bubbles of gas which would result in a decrease and final stoppage of the current, an effect known as "polarization."

THE DANIELL CELL

The Daniell cell, Fig. 4, is composed of a glass or stone jar containing a cylindrical plate of sheet copper and an inner porous pot or jar of unglazed earthenware containing a zinc rod. The porous pot is filled with dilute sulphuric acid, while in the outer part the copper plate is immersed in a saturated solution of copper sulphate. The electromotive force of the Daniell cell is about 1 volt. The internal resistance is about 3 ohms.

The Daniell cell is recharged thus: Thoroughly clean all the parts and re-amalgamate the zincs. Re-amalgamation is for the purpose of freeing the surface from impurities and is thus accomplished: Take a flat earthenware dish a little larger than the zinc plate, and put in just enough water to cover it. Then slowly add sulphuric acid, one part to the ten of water—that is, if you have 10 oz. of water, add 1 oz. of sulphuric acid. Immerse the zinc in the diluted acid and pour over it a small quantity of mercury which is to be painted onto the zinc with a little brush or mop made up of tow and a few brass wires. See that the zinc is well covered. The acid may be afterward used with the charging solution, and the mercury that is left may be poured back into the bottle.

After the zincs are thus re-amalgamated, charge the porous pot with a solution of 1 part of sulphuric acid to 12 of water, and the outer jar with a saturated solution of copper sulphate. On a sieve suspended in the liquid, place crystals of copper sulphate, to keep the solution saturated. Do not throw a handful of the sulphate into the jar to settle at the bottom. Care should be taken to keep the zinc and the zinc sediments from touching the porous pot, as this would attract particles of copper from the copper solution to form on the outer wall of the partition and cause a higher internal resistance and short circuit.

If properly understood and kept in working order, the Daniell battery is very constant, and is well suited to silverplating.

THE BUNSEN CELL

The Bunsen cell, Fig. 5, has several forms, of which the English and the French are the best known. The English Bunsen is similar in appearance to the Daniell, with a cylinder of amalgamated zinc in place of copper, and a square of carbon in place of the zinc rod. The outer cell is charged with dilute sulphuric acid, 1 part of acid to 10 parts of water, and the inner cell with strong commercial nitric acid. Run the nitric acid into the porous cell until about three-quarters full, and fill the space between the porus cell and the outer jar with the dilute sulphuric acid. The French Bunsen has sulphuric acid in the porous cell with the carbon, producing a constant generator with a low voltage, or electromotive force. This cell gives 1.8 volts at starting, falling to 1.5 volts when the circuit is closed. This form is less troublesome to keep in working order than the English Bunsen and is free from noxious fumes.

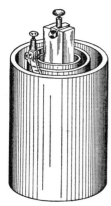

Fig. 5. Bunsen Cell

An English authority says: "The type of cell known as the Daniell is best for depositing copper from its sulphate solution and silver from the usual plating solution, but the Bunsen may be used for this purpose if the exciting liquid is sufficiently weakened. One or two large cells in series will be enough for all ordinary purposes" (presumably, for small work, such as the country repair shop).

The Bunsen cell is best for depositing copper and brass from their alkaline solutions, and also for the deposition of nickel, because its electromotive force is high, enabling it to pass through high resistances. It is not suitable for the work of silverplating, gilding, and electrotyping because its e.m.f. (electromotive force) causes the metal to go on too fast, and in a granular condition. In all these operations the Daniell cell will be found to be the best, because its e.m.f. is lower than that of the Bunsen, and its current equally constant in volume. The Smee cell is eminently useful for gilding small pieces of jewelry. Batteries with a high e.m.f. cause gold to go on too fast, and to give the deposits a brown color. So much has been said, though very briefly, descriptive of the leading types of batteries in use. The beginner will find it to his advantage to study the battery which he has adopted, and to become thoroughly familiar with its operation.

The cells forming the battery are connected up in various ways to obtain different results. We will explain, as briefly and simply as possible, the working of the cells in the battery, and how the electric current is controlled, since it is important in plating operations to get the proper pressure and current, depending upon a known resistance. Scientific terms will be avoided, except when necessary to make the statement exact.

In considering the flow of current in an electric circuit, we can compare it nicely with the flow of water in a water pipe. Suppose we had a pail of water with a pipe in the bottom. We know that the higher the water stands in the pail, the more rapidly the water will flow through the pipe, and that it makes no difference how big the pail, providing the water level is the same. Now we can think of a battery cell as a pail of water and an electric current as the flow through a pipe leading from the pail. To correspond to the pressure of the water on the bottom of the pail, we have the electromotive force which, in fact, is usually called the "pressure" or "voltage," and which we will call hereafter the e.m.f. The rate of flow of the electric current corresponds to the rate of flow of water, and it depends upon the pressure (e.m.f.) of the battery, just as the rate of the water in the pipe depends upon the pressure of the water in the same way. No matter how big the battery, the current will be no greater in the circuit if the pressure remains the same. We also know that in a water pipe, other things being the same, the larger the pipe, the more rapidly will the water flow (in gal. per min.). In the electric circuit, the larger the wire, the greater the

current will be, for the larger wire offers less resistance to
the flow of electricity. Some substances offer more resistance to the flow of electricity than do others; that is why
copper wire is used so much, as copper offers less resistance to the current than most other substances. Now if
we put a valve in our water pipe, we can change the flow
of the water (the current) by opening or closing the valve,
that is, we put more resistance in the path of the water.
In the same way, we can put a valve in the electric circuit,
by which we can vary the current from the battery by in-
troducing more or less resistance. Such a valve is called
a "resistance box" or "rheostat."

The pressure of water is measured in lb. per sq. in.;
electric pressure is measured in volts. The rate of flow
of water is measured in gal. per min.; the rate of flow of
electric current is measured in amperes; and the resistance
offered to this flow is measured in ohms. There are two
resistances to be considered, that of the cell or battery
itself; called "internal resistance," and that of the circuit
(the wires, solution, etc.) called "external resistance."

There are many ways of grouping the cells in a battery,
but they are comprehended in the systems known as
"series" and "parallel" and combinations of the two, called
"series-parallel."

In series, the cells are connected with the positive element of one joined to the negative of the next and so on,
and the free elements connected to the main circuit, as
shown at A, Fig. 6. In this arrangement, the current gen-
erated in the first cell has to flow through the second, and
the e.m.f. of one is added to that of the other. This method
corresponds to setting one pail of water on another, in
which case we double the pressure.

The parallel grouping of the cells has a different effect.
In this, all the negative elements are joined together on
one wire, and all the positive elements to the other wire,
as shown at B. This is called "joining for quantity." The

zincs are practically all one large zinc, and the coppers (or carbons) are one big positive element. The force is no greater, but the internal resistance is decreased in propor-

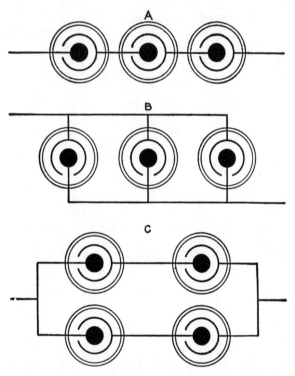

Fig. 6. Grouping of Cells—A, Series—B, Parallel—C, Series-Parallel

tion to the number of cells, as there are more paths through which the current may flow. The current is therefore increased in proportion to the number of cells thus connected. The e.m.f. is no more altered than would be the total water pressure, produced by placing two pails of water side by side on a level floor, in place of one; for both cells are giving the same pressure, and coupling them in parallel is only equivalent to increasing the size of a single cell, which has no influence on the e.m.f.

The series-parallel grouping, shown at C, unites two or more sets of cells which were joined in series in a parallel arrangement.

The practical results of these combinations may be figured out from a knowledge of the capacity of the cells and the resistance, by a formula known to electricians as Ohm's law, which declares that the current is equal to the e.m.f. divided by the resistance. This law can be expressed by a formula, which for any cell is $I = \dfrac{E}{r + R}$ where I equals the current in amperes, $E =$ the e.m.f. in volts, $R =$ the external resistances of the circuit in ohms, and r the internal resistance of the cell in ohms.

Suppose we have four cells of a voltage of 1.5, an external resistance in tanks, wires, etc., of 2 ohms, an internal resistance of .05 ohm, and wish to know the current given by the three combinations. Here $E = 1.5$, $r = .05$, $R = 2$. The various methods of grouping are as follows:

A.—SERIES GROUPING.

Electromotive force of one cell = 1.5; e.m.f. of 4 cells in series = 4 × 1.5 = 6 volts.

Internal resistance of one cell = 0.05 ohm; internal resistance of 4 cells = 4 × 0.05 = 0.2 ohm. External resistance = 2 ohms. Total resistance = internal resistance of 4 cells + external resistance = 0.2 + 2 = 2.2 ohms.

$$\text{Current (I)} = \frac{\text{electromotive force}}{\text{total resistance}}$$

$$\text{Current} = \frac{6}{2.2} = 2.73 \text{ amperes, nearly.}$$

B.—PARALLEL GROUPING.

Electromotive force of cells in parallel is the same as that of a single cell, if the cells are identical. That is, the e.m.f. = 1.5 volts.

The internal resistance of 4 cells in parallel is ¼ of that of one cell. That is, the internal resistance is 0.05 ÷ 4 = 0.0125 ohm. The external resistance is 2 ohms. Therefore, the total resistance = 0.0125 + 2 = 2.0125 ohms.

$$\text{Current (I)} = \frac{\text{electromotive force}}{\text{total resistance}}$$
$$= \frac{1.5}{2.0125} = 0.74 \text{ ampere}$$

C.—SERIES-PARALLEL GROUPING.

Electromotive force is double that of one cell; that is, the e.m.f. = 1.5 × 2 = 3 volts.

The internal resistance of each set of two cells in parallel is one-half that of each cell, = 0.05 ÷ 2 = 0.025 ohm. The total internal resistance of the two groups in series is twice that of either group singly, that is, 0.025 × 2 = 0.05 ohm. External resistance = 2 ohms. Therefore, total resistance = 0.05 + 2 = 2.05 ohms.

$$\text{Current (I)} = \frac{\text{electromotive force}}{\text{total resistance}}$$
$$= \frac{3}{2.05} = 1.46 \text{ amperes}$$

Suppose, however, that the external resistance is not so great as assumed in the above examples, in which it is purposely made rather high. On a "short circuit" (very little external resistance), $R = 0$ and we find the current from these three combinations as follows:

(A) $I = \dfrac{1.5 \times 4}{.05 \times 4} = \dfrac{6}{.2} = 30$ amperes.

(B) $I = \dfrac{1.5}{.05 \div 4} = \dfrac{1.5}{.0125} = 120$ amperes.

(C) $I = \dfrac{1.5 \times 2}{.05} = \dfrac{3}{.05} = 60$ amperes.

So that, while with a heavy external resistance we get the greatest volume of current from the series coupling, with a low resistance we get the greatest volume of current from the parallel, which at the same time gives the weakest voltage. From the above it is evident that the voltmeters and rheostats which will be treated later, must be sharply watched if the plater would know what he is doing.

THE DYNAMO

After learning the rudiments of the trade with such small electric batteries as are found most practical, the electroplater may feel warranted in using a dynamo. If one has some electrical training and mechanical ability he may be able to build a very serviceable dynamo for himself. This, however, is a separate trade, and the electroplater is apt to have his hands full in his own particular line. As a rule, it will pay him better to buy a good dynamo and use all his time in his regular business. The advantages of a dynamo in a shop where a good custom trade is established are that it yields a current in every respect more suitable to the work of depositing metals than from the best batteries, it is more cleanly in working, eliminates noxious gases, and is more easily managed.

The best form of dynamo for the use of the electroplater is some sort of shunt-wound or compound-wound machine, giving a large volume of current at low pressure, the current being delivered constantly and in one direction. In the shunt-wound dynamo, there is automatically a continuous balancing of the current as the load varies in the external circuit. The compound-wound machine, however, can also be designed to give a constant current, and has, in addition, self-regulating features not possessed by the shunt-wound machine. It is the best type of machine to use for electroplating. Such a generator is shown in Fig. 7. Machines for electric lighting work are designed to give a moderate volume of current at high pressure, and while a plating dynamo will generate current at, say, 10 volts, the lighting current may have a pressure of 110 or 220 volts, or higher. The unfitness of the lighting current for the purposes of the electroplater is therefore manifest.

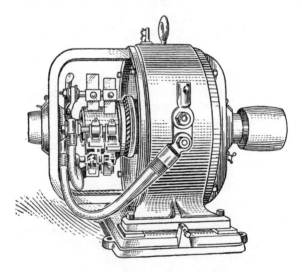

Fig. 7. Compound-Wound Dynamo

In purchasing a dynamo, avoid all offers of second-hand machines, although warranted "as good as new." Get a small machine of a responsible dealer, at a cost of $50 or upwards, and then master its construction and working so as to get the best results. It may be set down as a law in electrical engineering, though probably never before printed, that a dynamo will deteriorate and go to scrap with a velocity equal to the square of the ignorance of the operator.

REGULATING AND MEASURING INSTRUMENTS

A rheostat, to enable the plater to control the current, is necessary where a dynamo or large battery is used. One style of rheostat is shown in Fig. 8. Throwing aside technical terms, the rheostat is similar to a tap or valve in a

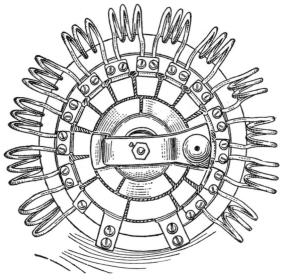

Fig. 8.   Rheostat

water pipe, each wire taken in by it acting like the plug of a tap narrowing the orifice through which the liquid flows.

The ammeter, Fig. 9, measures the rate of flow of current, in amperes, and is sometimes made in connection with the voltmeter. As 1 ampere of current, under proper conditions, will deposit in 1 hour 62.1 gr. of silver, 18.3 gr. of copper (from cuprous solutions), or 36.6 gr. of gold (from aurous solutions), it is evident that the plater should regulate his current according to the work in hand.

The voltmeter is similar to the ammeter, but measures the pressure of the current in volts. In Fig. 10 is shown

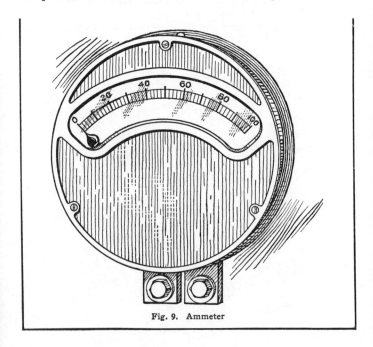

Fig. 9. Ammeter

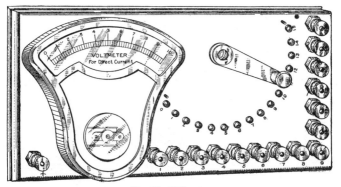

Fig. 10. Voltmeter

a voltmeter with a switch for connecting it to any given number of tanks. Experience shows that the silver in a plating solution may be separated from its salt and deposited in good condition with a current at as low a pressure as 2 volts, which may be increased, if required to work more rapidly, to even 3 and 4 volts. When the pressure exceeds the latter figure, however, there is a tendency to a loose and powdery deposit. So it becomes necessary to regulate the pressure, or voltage, as well as the current or number of amperes. It is not necessary to explain here the construction of these meters, or even their arrangement, the latter depending on the number and size of the tanks and batteries used, and full directions may be obtained of the supply house from which the equipment is purchased.

CHAPTER III

SHOP EQUIPMENT

W E WILL NOW consider the tanks, scrubbing trough,
polishing lathes, brushes and other furniture.
Trough and tanks may be made by the beginner himself, if
he is a good mechanic. But if he is not handy with tools, or

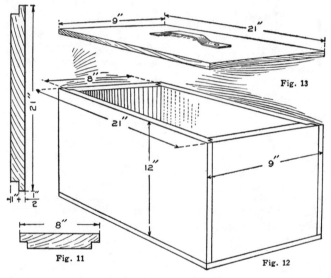

Fig. 13

Fig. 11

Fig. 12

Homemade Electroplating Tank

if his time can be profitably employed otherwise, he would
better buy them ready made, or have a carpenter make
them for him.

To make a small tank (inside dimensions 6 by 18 in.
with an inside depth of 12 in.) you will require—

2 pieces for sides 12 by 21 in.
2 pieces for ends 8 by 12 in.
1 piece for bottom 9 by 21 in.

The wood should be 1½ in. thick and of good soft wood without knots. The pieces must be cut true and square, or it will be impossible to make the tank water-tight. Rabbet the ends of every piece except the 9 by 21, which is the bottom. Fig. 11 shows the rabbeting of the boards and Fig. 12 shows how they are put together. Use large wood screws, not nails, in joining the parts. Before joining, brush white lead thickly where the grooved ends come together, and put in the screws while the white lead is still wet. Then take some cotton waste or like material and calk all the crevices. Calk from the outside of the tank.

The next process is to line the tank with asphaltum. For a tank of this size, allowing 1 lb. for each gal. of capacity, 6 lb. of asphaltum will be required, the capacity of this tank being just 5.6 gal. This will be a liberal allowance and make a good tank lining. The asphaltum should be secured from some good platers' supply house, because if mixed by an inexperienced person it is liable to get hard and crack. After lining the inside of the tank with asphaltum, the outside may be painted, or it may be brushed over with ordinary asphaltum.

Covers for tanks, which are necessary to keep out dust, etc., may be made of any ordinary wood, and should be the exact width and length of the tanks. It is well to brush the inside of the lid with the prepared asphaltum. Fig. 13 shows the cover for the tank.

To suit various sizes of work it may be necessary to make the tanks the width of two planks, in which case they should be matched tightly and drawn together with four rods, two at each end of the tank, with a screw and nut on the ends.

Plating vats may be bought ready made, the 10-gal. size

costing about $10.00. A 15-gal. scrubbing tank costs $5.00 or $6.00.

For goldplating, or gilding, a crock or glazed dish may be used. One holding a pint of solution is usually large enough, as the filling of a large tank with gold solution would, of course, be out of the question. Another reason for using the crock is that the gold solution is used hot.

The plating tank is made ready for work in the following manner: After testing the tank carefully for leaks, and calking and brushing in more asphaltum if necessary, place it on wooden horses in the place assigned to it, of

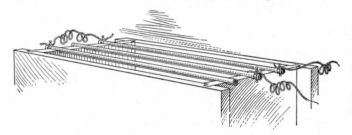

Fig. 14. Top of Tank with Pipes in Place

which more will be said when we come to the subject of shop arrangement. See that the tank is plumb and steady and not to be disturbed by jostling the supports. Get three pieces of brass pipe, ¼ or ⅜ in. inside diameter, for each tank. They must be stiff enough to hold the anodes and the articles to be plated. Set these pipes in grooves in the ends of the tank, as shown in Fig. 14, and connect with copper wire, carefully soldered, or connected by brass or copper clamps.

The practice of winding the wire about the ends of the rods without soldering is a bad one, as the wires are apt to get loose and the winding filled with dirt, making a bad connection. See that all joints are well made, and solders and all parts connected by clamps and under screws are made clean, with broad surfaces of clean metal in contact at all points. Keep the metal clean by occasionally scouring it with emery paper. Perfect connections insure an uninterrupted circuit and the highest economy in working.

The arrangement of the connecting rods and the tanks is shown in Fig. 15. On the positive rods AA are hung the anodes, composed of the metal to be dissolved by the action of the electric current, and on the negative rod, B, are suspended, by hooks of No. 8 gauge copper wire, or by the regular slinging hooks that come for this purpose, the goods to be plated, which are the cathodes in the electric

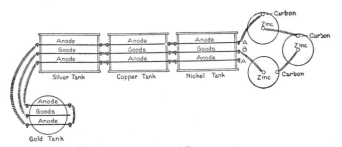

Fig. 15. **Arrangement of Tanks and Battery**

circuit. This figure also shows how four tanks may be arranged for different classes of work and connected with three or more cells as may be needed. When a single tank only is to be used, connect that tank only with the battery. It should be understood, of course, that a small job requires less current in amperes than a large one, and that all current used in excess of needs is not only a waste, but injurious to the job on hand.

The tanks should never be left empty when not in use, nor should the solutions be allowed to get foul, for if the tanks get dry they are liable to warp and leak, and dust and impurities in the solution will interfere with the plating.

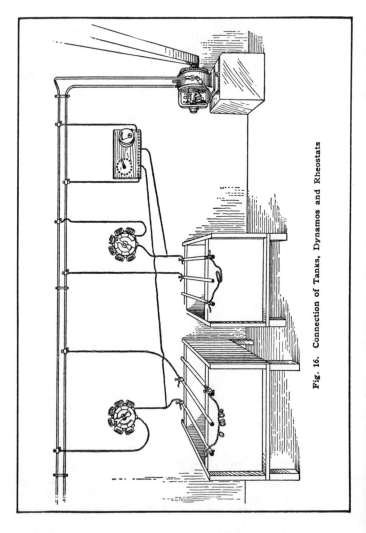

Fig. 16. Connection of Tanks, Dynamos and Rheostats

TANK REGULATION

Tank regulation is an important matter in an up-to-date shop where several tanks are in constant use. There is a material difference in the density of different solutions, which varies also as they are used, and in manipulating tanks of various sizes where different solutions and different operations are employed it becomes necessary that proper regulations of the current be secured. This cannot be effected by the field rheostat on the dynamo, which controls the voltage at the terminals of the machine by increasing and decreasing the resistance provided by the circuit, but does not affect the ampere capacity of the machine. On the other hand a rheostat placed in series with the tank affects both amperes and volts, reducing the latter in the proportion in which the former is cut down. Therefore in selecting a tank rheostat it should be of a size suited to the ampere capacity of the tank. Fig. 16 shows the connection of two rheostats in the circuit for tank regulation. In this case a dynamo is used instead of a battery to supply the power.

SCOURING TROUGHS

The scrubbing or scouring trough may, of course, be made large or small, according to the size of the shop. For convenience this is made in shape like the ordinary stationary wash tub, and should be of heavy dressed plank, well matched and jointed. As strong potash solutions are used in scrubbing, the trough or tray is sometimes painted with prepared asphaltum, or lined with lead, the seams of which must be wiped and not soldered.

Rabbet the bottom and sides, and bevel the front side to give the required angle. Set the bottom into the sides, not the sides on to the bottom. Fit the joints together tightly with white lead. Use large wood screws, not nails, in putting the parts together. Set a couple of horizontal

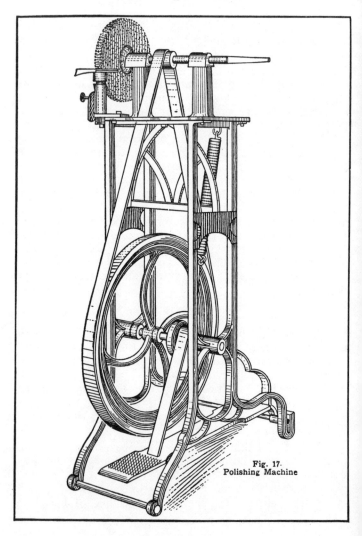

Fig. 17.
Polishing Machine

cleats 6 in. down in one end of the trough and make a light tray on which to rest goods in working.

MISCELLANEOUS EQUIPMENT

A treadle lathe, scratchbrush and a bench polishing ma-

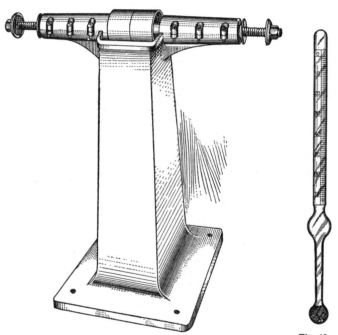

Fig. 18. Power Polishing and Buffing Lathe

Fig. 19.
Hydrometer

chine are indispensable. In a large shop these machines are run by power, and the plater must be his own judge as to the economy of using electricity or steam in running his tools. For a small shop, it is taken for granted that

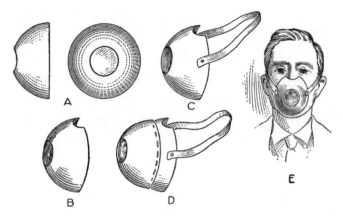

Fig. 20. Making a Respirator

one or two simple foot lathes will suffice to begin with, and these may be bought second-hand or even built by the expert mechanic. A simple form of foot-power polishing machine is shown in Fig. 17, and a power polishing and buffing lathe is shown in Fig. 18.

The cleaning, polishing and buffing wheels required are the following: Felt wheel; leather wheel; emery or carborundum wheel; cotton and flannel buffs; cotton watchcase buff; long, thin felt buff for the inside of finger rings, and brass-wire scratchwheel.

Among the miscellaneous equipment may be mentioned: Good scales, a clock, several galvanized iron pails, wooden buckets or tubs, various wood-fiber brushes, a glass funnel and graduate glass, thermometer, hydrometer, Fig. 19, (for measuring the specific gravity of the solutions), glue pot, etc. In the way of personal equipment, perhaps a suit of old clothes should not be omitted, including a stout pair of high shoes. Some platers use shoes with wooden soles, but the floor should never be so wet as to require such a precaution. A good rubber apron, rubber gloves and finger tips (the tips should not take the place of the gloves entirely), a respirator and a pair of automobile goggles complete the outfit. Remember that what may seem a useless precaution may save a man from being laid up for a week, or from more serious harm.

A serviceable respirator to guard the nose and throat from poisonous acids and dust, may be made as follows:

Get a thin rubber ball about 6 in. in diameter, cut it in half, and then cut out a circle 2 in. in diameter in each half as in A, Fig. 20, and a semicircular notch in the edge ¾ in. deep,

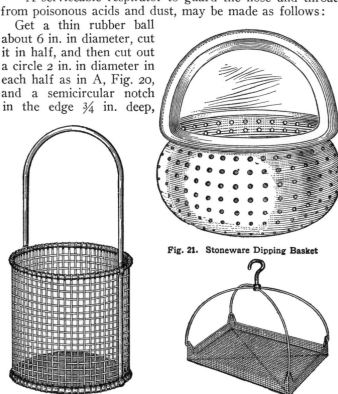

Fig. 21. Stoneware Dipping Basket

Fig. 22. Aluminum Dipping Baskets

as shown at the top in Fig. B.   Fasten onto one of
the halves a rubber band to go snugly but not tightly over
the head, as at C.   Sew the other half of the ball at the
sides, which allows a space to insert a close-grained sponge
as shown at D.   In use its appearance is shown at E.
When the sponge is dampened one may breathe easily
(through the nose always).   The sponge may be easily removed and cleaned.   See that it is always kept moistened.

A number of stoneware crocks and kettles will be found
necessary, and these are to be preferred to enameled ware,
as the enamel wears off in a short time and the iron is then
attacked by the acids.   A stoneware pitcher or two will
also be needed.

Dipping baskets of glazed stoneware are used where a
number of small articles have to be dipped in acid or
cyanide solution.   Aluminum wire baskets are now made,
and are much lighter, but they cost four or five times as
much as the earthenware, which run from 50c to $2.00.   A
stoneware dipping basket is illustrated in Fig. 21, and
aluminum baskets in Fig. 22.

THE PLATING SHOP

We now come to the plating shop and its arrangement.
Just as a kitchen may be comprehended in a chafing dish
or a hotel outfit, so the plating shop may begin with a
single-cell battery and a glass jar or stoneware kettle, and
grow up to an establishment covering an acre of ground.

We will suppose that the beginner starts with a small
shop.   This should be in the business part of the city, but
not where there is much smoke.   A good business street in
a growing suburb is a good location, for, as an old silverplater wisely says, "the trade should go to the people, and
not expect the people to come to the trade."   The shop
should have two rooms—one for the plating room and one
for the polishing room.   If there is a small room for an
office, so much the better.   One large room might be used
by dividing it with a partition.   The solutions, tanks, etc.,

must be carefully protected from dust, and this cannot be done if the polishing and cleaning are done in the same room with the tanks.

Light and ventilation are prime requisites. Artificial light will not supply the place of diffused sunlight. A northern exposure is to be preferred, and a good skylight is a great advantage. The examination of the goods and the solutions, while the various processes of the work are going on, and the care of the machines, to say nothing of the processes of polishing, etc., cannot well be undertaken without plenty of sunlight. At the same time care must be taken to protect the tanks from the direct sunlight, which decomposes all plating solutions. This is why a northern exposure is preferred. The vats and baths should be covered with canvass or wooden covers when not in use, to exclude both dust and light.

The shop should be well ventilated, through an ample opening in a skylight or by a fan in the chimney or air shaft. The gases and fumes of some of the chemicals used are highly poisonous, and care should be taken that they are carried off rapidly. Have a closet for the stock of acids and chemicals and keep them there properly labeled, under lock and key. In another cupboard keep antidotes for poisons, as given in detail in Chapter X. Handy cabinets, in which tools may be kept, may be made from soap boxes or other boxes.

The floor should be tight and smooth, and kept dry. There is no need of having a plating shop wet and sloppy, and if there should be tenants below, the drip might cause trouble. Use plenty of sawdust, and never let pools of water stand on the floor. When sweeping, be sure to cover all plating tanks, acid baths, etc., and use damp sawdust to gather the dust. Steam heat is almost a necessity on account of its regularity and freedom from dust, gas and smoke. The temperature of the shop should be 60° to 65° F.

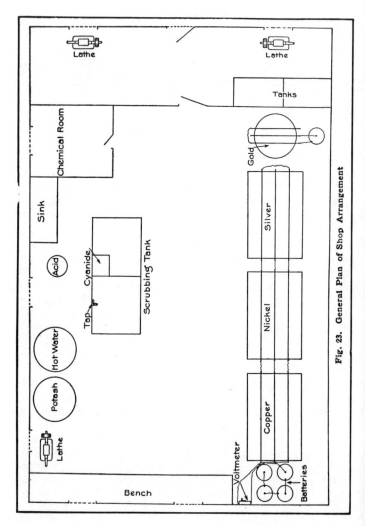

Fig. 23. General Plan of Shop Arrangement

Fig. 24. Circular Scratchbrushes

In arranging the shop it is well to get the aid of an experienced plater in whom confidence can be placed. No directions can be given in print that would cover all conditions of location of the various machines and apparatus, the arrangement of light, and the amount of equipment to be carried. A general plan of shop arrangement is shown in Fig. 23.

POLISHING TOOLS AND POWDERS

Scratchbrushes are made of both steel and brass wire and are used on a lathe for the first process of cleaning up. The stiff brushes may be used with emery, carborundum and other abrasives to grind off old silver and smooth the

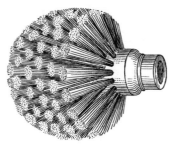
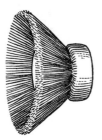

Fig. 25. Lathe Brush for Insides of Goblets

Fig. 26. Cup-Shaped Brush for Watch Cases

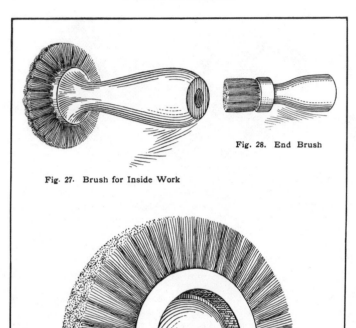

Fig. 28. End Brush

Fig. 27. Brush for Inside Work

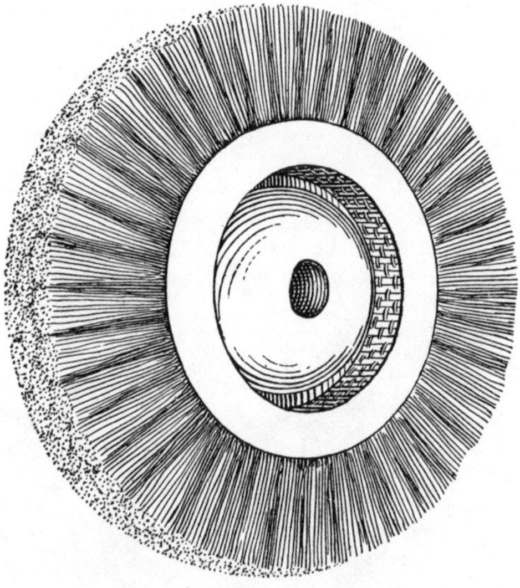

Fig. 29. Tampico-Fiber Brush

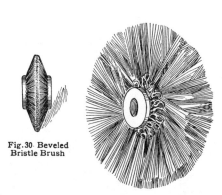

Fig. 30 Beveled
Bristle Brush

Fig. 31. Special Loose-Wire Brush
for Satin Finishing, Etc.

Fig. 32. Hand Brush

Fig. 33. Hand Brush

Fig. 34. Hand Brush

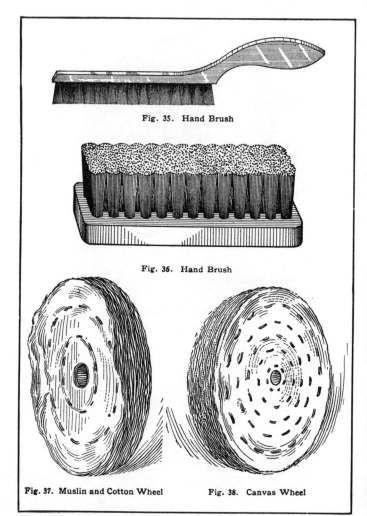

Fig. 35. Hand Brush

Fig. 36. Hand Brush

Fig. 37. Muslin and Cotton Wheel

Fig. 38. Canvas Wheel

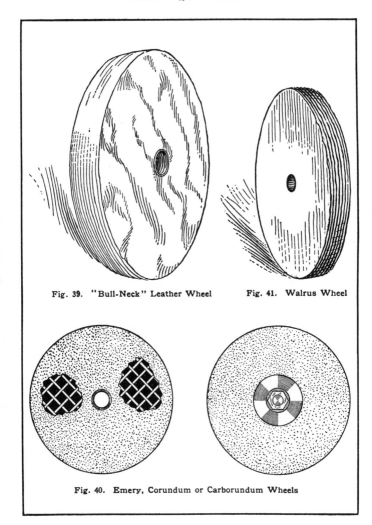

Fig. 39. "Bull-Neck" Leather Wheel   Fig. 41. Walrus Wheel

Fig. 40. Emery, Corundum or Carborundum Wheels

rough surfaces of castings, while the finer brushes of crimped wire may be used for gold work.

Figure 24 shows two styles of platers' circular scratchbrushes for work on the lathe. A lathe brush for the inside of goblets is shown in Fig. 25, while Fig. 26 shows a cup-shaped brush for watch cases. Another brush for inside work is shown in Fig. 27, and an end brush in Fig. 28. A tampico-fiber brush for the lathe is shown in Fig. 29, and a beveled bristle brush in Fig. 30.

In order to produce the satin finish or sand-blast effect on German silver, aluminum, brass, steel or other metal, special brushes are used. As will be seen in Fig. 31, the wires are loosely hung to the hub and are allowed to whip the work. The common types of hand brushes are illustrated in Figs. 32, 33, 34, 35 and 36.

Buffs or bobs may be used on the same lathe as the scratchbrush or emery wheel, on the right-hand end of the spindle. They are used both for polishing new work and for finishing plated goods. Buffs are in endless variety, of felt, leather, canvas and muslin. The usual style of stitched muslin and cotton wheel is shown in Fig. 37. When the term "calico" is used in English or German technical b o o k s, unbleached muslin is r e f e r r e d to. Swansdown calico is a soft, unbleached muslin. Figure 38 illustrates the c a n v a s wheel. A "bull-neck" leather wheel is shown in Fig. 39.

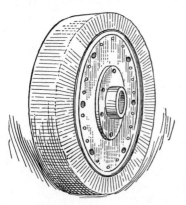

Fig. 42.   Compressed-Leather Wheel

Emery, corundum, or carborundum wheels, illustrated in Fig. 40, are used on the lathe for grinding down new goods, removing old plating, etc. The use of the

Fig. 43. Correct Method of Holding Article in Polishing or Buffing

emery wheel is usually followed by the bull-neck leather wheel, walrus, and canvas.

Walrus wheels are made of tanned walrus or seahorse hide, which from its extreme toughness is regarded as indispensable in certain classes of work. Such a wheel is shown in Fig. 41, and the compressed-leather wheel in Fig. 42. The right method of holding the article when polishing or buffing is illustrated in Fig. 43. Let the wheel run so that the bottom of the wheel goes from you.

Pumice, powdered, is used with a scrub brush and lye solution to clean iron and steel goods and to remove dirt and grease. The best pumice comes from Italy.

Emery or carborundum is used in polishing old tableware. It is glued on leather or felt wheels to grind out rough places in all kinds of work. Turkish emery is generally used by electroplaters, and an emery glue may be bought already prepared.

Silver sand is used to remove the scum in silverplating in the same way that the wire scratchbrush is used. Places that cannot be reached with the scratchbrush, as the inside of rings, cups and hollow-ware, are treated with wet sand on a brush or bob.

Tripoli is a fine siliceous earth which is made into a grease composition. It is used on all kinds of wheels except cotton flannel, bristle and wire brushes, and is good for cutting down and smoothing all kinds of metalwork. To apply tripoli, hold it against the wheel while the latter is running.

Rouge is a red powder and is practically the same as iron rust which has been heated to drive off the water that is combined with it. It is applied to wheels to brighten and color up all kinds of work, both before and after plating. It is used in the same way as tripoli. French rouge is the best.

Crocus is a deep yellow or a red powder, the oxide of some metal (usually iron) which has been calcined at a great heat. It is a little coarser than rouge, and is used in the same way.

A pure, soft lime, free from grit, is used as a polisher. Vienna lime is used in this country and Sheffield lime in England. It comes in 15-lb. cans.

Litmus paper should always be kept on hand for testing solutions. Blue litmus turns red if the solution contains free acid, and red litmus will turn blue if free alkali is present. Keep it in a tight tin box.

Whiting is chalk, ground to a fine powder. It is an excellent polisher, used after pumice.

Rotten stone is a decomposed silica derived from certain

limestones, and is used as a polisher. The best comes from England and is as fine as flour.

MECHANICAL PLATING

Where a large number of small articles are to be electroplated, the time required in stringing them on wires for hanging in the tank becomes a considerable item. To do away with this expense, a special tank has been brought out. This tank has a perforated cylinder in which a large number of small articles may be placed and which is kept in constant rotation by a belt.

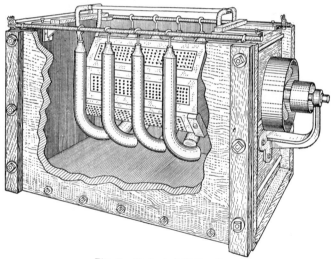

Fig. 44. Mechanical Plating Tank

The construction of this mechanical plating tank is shown in Fig. 44. The articles to be plated are poured into the cylinder as shown in Fig. 45, and the anodes of copper, nickel, or whatever the coating is to be, are hung in parallel rows on each side of the cylinder. Special curved anodes are used, making the operation quick and effective. The cylinder should be rotated at from 10 to 20 revolutions per minute, two speeds usually being provided for by a stepped pulley outside the tank.

Several advantages are claimed for these "plating barrels," and the results seem to bear out the claims made for them. The principal feature is, of course, the labor saved in preparing the articles. This is strikingly shown by a comparison of Fig. 45 with Fig. 46. In plating bronze or

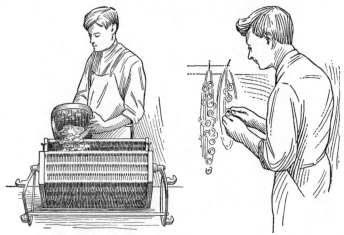

Fig. 45.　Filling the Plating Barrel—
The New Way

Fig. 46.　Preparing Small Articles for
Plating—The Old Way

brass on steel in stationary tanks, there is a tendency to a deficiency or an excess of the copper, but this seems to be overcome with the mechanical platers. Further, the deposit of metal seems to give a more even and more protective coating with a somewhat smoother finish.

Any small work, such as small automobile and bicycle parts, bolts, nuts, screws, sewing-machine and typewriter parts, and the like, are particularly adapted for mechanical plating.

CHAPER IV

CLEANING GOODS BEFORE PLATING

WHEN GOODS are received, make a careful tally, giving kind, number, marks, etc., with directions as to plating. The first operation is the cleaning of the goods. In plating, it should be carefully understood that "cleaning" means making the goods chemically clean, not ordinarily clean, as an entirely unobstructed surface of metal must be exposed to the solution to insure the thorough adhesion of the particles of metal to the cathodes (or goods to be plated in the tank). It is necessary, therefore, to free the surface of the goods, including every line and indentation, from every trace of foreign matter of any kind, whether in the form of rust, verdigris, tarnish, or any other kind of corrosion, or in the form of oil, grease, lacquer, sweat, etc. The touch of a finger on the prepared surface is sufficient, for instance, in silverplating, to cause the silver to strip off from the spot when the scratchbrush or the burnisher is applied.

The workman must also see that all deep scratches, dents and cracks are removed before the goods are put in the tanks, and that all necessary repairs are made, avoiding an excess of solder, and that all previous coats of silver or nickel are removed and the surface polished.

CLEANSING SOLUTIONS

To remove grease, etc., first plunge the goods in hot lye, using ½ lb. of caustic potash or soda to 2 gal. of water; then plunge them in hot water and give a final rinsing in cold water, when they are ready to be treated for rust and corrosions. If there is no steam coil, the lye and water kettles may be firmly set over a gas furnace.

Rust, verdigris and other metal oxides are removed by a

53

pickle of a dilute mineral acid. A mixture of equal parts of sulphuric acid and water makes a good cleaning bath for these oxides. String the articles on a wire and plunge them in the mixture for a few minutes. Use the scratch-brush and wash, and if the stain is not removed, repeat the process.

A light tarnish may be removed by dipping in a strong solution of cyanide of potassium, say, 10 oz. of cyanide to 1 gal. of water, with the addition of a few drops of liquid ammonia. For the oxides of copper and zinc, use a mixture of 1 part of sulphuric acid in 20 parts of water, and if the stains are deep and fail to come off, strengthen the pickle.

For rust on iron or steel, use a pickle of 6 parts sulphuric acid, 1 part muriatic acid, and 160 parts of water. The goods are dipped in this solution from 15 to 30 seconds. For oxides of lead and tin, including britannia ware, pewter, etc., use a hot lye solution.

Lacquered goods should be steeped for an hour or so in warm methylated spirits and then immersed in a strong solution of ammonia. They should then be well brushed with an old scratchbrush and rinsed in hot water.

When these cleaning baths are not in use, they should be put up in jugs of vitrified stoneware, carefully marked, and put away in the chemical closet. The mixed acids will eat the glaze of ordinary earthenware as well as the enamel from iron.

The next process after cleaning, if the job is one of re-plating silver, is to remove all the old silver plate. For this purpose a stripping solution is used.

STRIPPING SOLUTION

To one gallon of commercial sulphuric acid, add 12 oz. of potassium nitrate (saltpeter) and ½ teaspoonful of French rouge. Do not put soft metal, such as britannia, lead, pewter, tin, zinc, etc., in this solution, or any wet articles, as they would be eaten by the acid. Fifteen minutes is usually long enough to leave goods in this pickle. After the old plating is removed, the surfaces of the articles to be plated should be well brushed in water with a stiff brush to remove loosened dirt from crevices. This process may reveal several imperfections in the way of dents and scratches which should be taken out on the polishing lathe or by the careful use of the hammer. After polishing, the goods should be again plunged into the alkali pickle to remove the film of grease which was put on in handling. The dipping, however, adds a thin coat of black oxide, which must be removed by scouring. For this purpose the article is held on the dipping tray with the left hand, while the article is brushed with a wet brush dipped in pumice powder until every trace of oxide is removed. Then rinse in water, and it will be quite clean and ready for the plating tank.

The article should not be touched again with the hand, but should be picked up with a wire hook, and, if to be silverplated, dipped into the necessary bath, again rinsed in clear water, and put into the plating tank.

If the goods are of iron, steel or zinc, they should be copperplated before the silver is put on. Many platers do this with soft-metal goods also, getting a better job by this process.

CHAPTER V

THE PROCESS of copperplating will be treated first, as it is a preliminary to much of the nickel, silver and gold plating, for the reason that these metals, especially silver, deposit more easily and firmly on copper than on iron, steel or soft metals, which take copper easily and then present a good surface for the finishing metal. Again, the preliminary coppering enables one to see whether the goods are properly cleaned, as the copper will strip off from greasy and corroded spots under the scratchbrush, while scratches and pinholes will be well filled. When rough goods are well polished and coppered, they will turn out a good job in the nickel or silverplating tank. If repairs are necessary, they should be smoothly made with a chloride-of-zinc flux applied with a copper wire, giving a chloride-of-copper amalgam which takes the copper readily.

When the goods have been thoroughly cleaned and polished, as previously described, they are hung on the suspending wires, plunged in the potash pickle and transferred to the plating vat without rinsing. The copper solution for the plating tank is prepared as follows:

COPPERING SOLUTION

Dissolve in an iron vessel 2 lb. of copper sulphate to 1 gal. of hot distilled water. Let it cool, and add liquid ammonia slowly to precipitate the copper, which takes the form of a green mud. Then add more ammonia, stirring with a stick until the mud is dissolved and the liquid takes a bright blue tinge. Use about 4 oz. of liquid ammonia. Now add gradually 12 oz. of potassium-cyanide solution, which will turn the solution to a dull amber tint. Leave exposed to the air for 12 hours, filter through two thicknesses of muslin, and dilute with 3 gal. of distilled water. The solution, after standing a day, may be worked cold, but it is better to raise it to 150° F. The solution is strengthened from time to time by adding ½ oz. or so of cyanide of potassium, carefully stirred in, and the same amount of ammonia.

For the anode plates, pure copper is employed, with a surface somewhat in excess of the surface of the goods to be plated. Too much anode surface is likely to give a hard, dark deposit, especially if the current is high, while too little anode surface is apt to give a loose deposit, which will peel off under the scratchbrush. Too rapid a deposit is to be avoided.

In placing the articles in the vat, or rather when rinsing them in the potash pickle, they are hung on a copper-wire "S." See that all bubbles are washed off when hanging the goods in the vat. If the goods are gently moved during the plating, the deposit is made brighter.

Ten or fifteen minutes is usually long enough to secure a good plating. Take out the goods, rinse in hot water, and dry in hot sawdust. They are then ready for polishing.

CHAPTER VI

F OR NICKELPLATING, the goods are prepared as for any other work, and if treated with a thin coat of copper, as described in the last chapter, the nickelplating will be brighter and more durable.

Be very particular about having the goods properly dipped before putting them through the plating bath. As they come from the polishing lathe, although apparently bright, they are covered with a thin film of grease, invisible to the eye, but sufficient to prevent the nickel in the solution making contact with the metal surface of the goods. This would cause blister, which would show in the scratch-brushing. To avoid this, rinse the goods well in the lye and scour with powdered pumice stone or whiting, for fine work, holding them with a dry linen towel well powdered. When well scoured, rinse off the powder or whiting in clear water, plunge the articles in the muriatic-acid dip described elsewhere, rinse and hang in the copperplating vat for the preliminary coating of copper, or place directly in the nickel vat if the coppering is to be dispensed with. If the goods have been coppered, examine the work under a good light, rinse and place in the nickel vat.

It is of no advantage whatever, but rather a disadvantage, if the goods have been previously nickelplated, for all the old nickel must be stripped before plating with a new coat. This is due to the peculiar action of nickel in the plating solution, as it refuses to make a permanent union with a nickel surface, even one newly plated and polished. If the goods are in the form of tubes or plain flat surfaces, the nickel may be taken off with an emery wheel; otherwise it must be put into the acid stripping bath.

NICKEL-STRIPPING BATH

In an iron pot or lead-lined tank holding 4 gal., put ½ gal. of water, and add slowly and carefully 2 gal. of strong sulphuric acid, stirring with a smooth hardwood stick. Be careful not to let it spatter out on the hands or face, as the

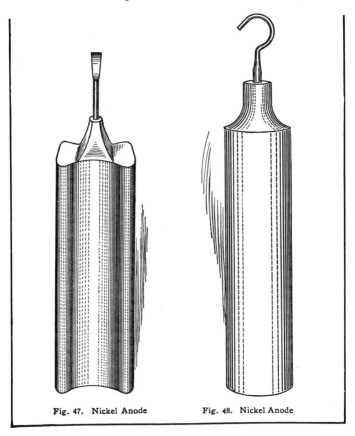

Fig. 47. Nickel Anode   Fig. 48. Nickel Anode

acid is very corrosive, and on contact with the water is raised to a scalding temperature. It is best to wear a mask and rubber gloves when handling this or other acids. When the sulphuric acid is mixed with the water, add ½ gal. of commercial nitric acid, and stir as before. When the mixture is cool, pour it into a stoneware jar, which should be protected by a good cover.

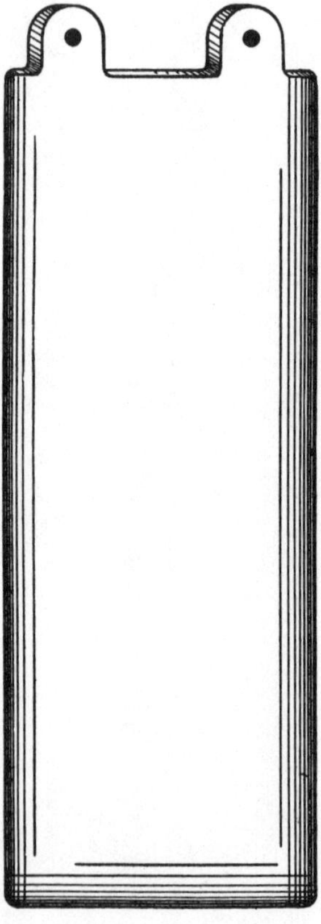

To strip nickel goods, clean them properly as before directed, plunge them in the alkali solution and then in the hot-water tank, and, by means of a copper wire, drop them in the stripping bath. The process will require five minutes to half an hour, according to the amount of nickel to be removed. The goods can be drawn up and inspected as often as necessary to see how the operation is proceeding. As soon as the nickel is entirely stripped, take out the goods and rinse in clear cold water, again in hot water, and dry in hot sawdust.

NICKEL SOLUTION

Dissolve double sulphate of nickel and ammonia in the proportion of ¾ lb. to the gallon of distilled water. Use an iron kettle and have the water boiling. When cool, filter through muslin and pour into the vat. Be particular to use only the

Fig. 49. Nickel Anode

best quality of nickel salts, which give a brilliant, silver-like effect. Care should be taken to use only pure nickel for the anodes. Several special shapes of nickel anodes are shown in Figs. 47, 48, 49, 50 and 51. In placing the work in the vat, put in the larger articles first, if there are several sizes, and do not attempt to plate different metals, as steel and britannia, in the same tank,

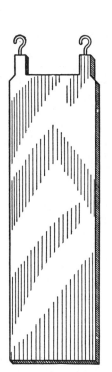

Fig. 50. Nickel Anode

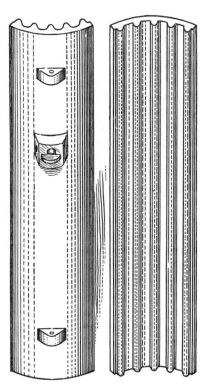

Fig. 51. Nickel Anode

as these attract the nickel in varying degrees. See that sufficient anode surface is used, else the solution becomes too acid. For ironwork, a slight acidity does no harm, and is rather a benefit. The degree of acidity may be tested by litmus paper, an acid solution turning blue litmus paper to red and an alkaline solution turning red litmus paper to blue. A special hydrometer, called a "nickelometer" is used by nickelplaters to test the solution. This solution should not register less than 7 on the nickelometer.

To correct the excess acidity, add liquid ammonia in small quantities until the solution ceases to redden blue litmus paper. Excess of alkali is shown by a yellow deposit and by turning red litmus paper to blue. In this case, use sulphuric acid until a slight acid reaction is shown. If the solution becomes weak and fails to deposit white nickel, add common salt at the rate of 50 to 70 gr. to the gal. and increase the anode surface.

The length of time in which the goods are to remain in the vat depends upon conditions. If previously coppered, not so long a time is required, and, similarly, if the solution is strong and there is a good current and sufficient anode surface, less time is required. Nickel is deposited at the rate of 16.89 grains per ampere-hour. For a single tank, a current supplied by a battery of 3 or 4 cells in series would be required. Watch the current, and after a thin coating of metal has been deposited, reduce the voltage by one-half or more. Leave the goods in the bath 2 or 3 hours, according to the class of work to be done.

When the goods have taken on the desired thickness of metal, lift them out of the solution by the slinging wires, rinse in the hot-water tank and dry off quickly in hot sawdust. Work quickly and be sure the rinsing and drying are thoroughly done, as exposure to the air with the plating solution causes oxidation in the form of blotches which polishing cannot remove. The goods, as taken from the vat, have a dull gray appearance, and may range from a creamy white to a dull yellowish hue. A dirty gray indicates an inferior solution, or perhaps a too rapid deposit in the bath.

After thorough drying in sawdust, the goods are well brushed with soft, clean rags dusted with Vienna white (purified chalk), tripoli and rouge. This may be done either by hand or on the polishing lathe, using soft muslin rags, and a small brush to reach the crevices. Do not use the brass scratchbrush, as the brass will make a deposit on the nickel which is difficult to remove.

Imperfections in the plating, such as an occasional blister, may be removed by sponging the spot with the alkali dip, to cleanse it, and applying a sponge wet with the nickel solution and containing a piece of nickel anode connected with the battery. Connect the goods with the negative pole of the battery as if it were in the plating tank, and let the current pass through for as long a time as if the goods were in the tank. This process can only be resorted to for an occasional patch. If the whole work is imperfect, it will have to be stripped and replated.

The plater cannot be too particular in finishing the goods and securing a high polish, as this is what chiefly gives satisfaction to the customer, who can only judge of the wearing quality of the work by experience. A variety of polishing brushes and mops should be used, and all roughness and clouds removed. A basil-leather mop (bark-tanned sheepskin) with fine tripoli imparts a fine polish, if finished with swansdown muslin and a little rouge. Much rouge is objectionable on account of the dust. Go over all the surface thoroughly, reaching all the holes and angles, and secure a bright, uniform polish. Beware of abrasives and too much "elbow grease" as such treatment is apt to cut through the nickel skin and ruin the job. The process of polishing emphasizes the importance of a good thick coat of nickel in the plating bath (and the same is true of all plating processes), as it allows for more effective finishing without spoiling the job and necessitating stripping and re-plating, with consequent expense and loss of time.

Nickelplated work after exposure to the air for a few months, especially in a damp climate, is apt to tarnish, causing dissatisfaction to the customer. The tarnish is caused by the oxidation of the minute particles of iron contained in the nickel, the iron being added to the metal in the anode for the purpose of "softening" it or making it more soluble. Good nickel contains from $\frac{1}{4}$ per cent to 2 per cent of iron, and the deposit is all the whiter on this account.

To prevent the discoloration in fine goods they may be run through an acid pickle to remove the iron on the surface. The pickle is made by adding 1 gal. of muriatic acid to 4 gal. of water. Sulphuric acid may be used, but it is not a good solvent. After the articles are plated, rinse in cold water and plunge them a moment in the pickle, which is warmed to 80° to 100° F. Let them remain a few seconds, rinse in cold water, then in hot water, and finally soak in a hot whale-oil soap solution for 15 or 20 minutes.

CHAPTER VII

SILVERPLATING is the rock on which many an amateur has been shipwrecked. Misled by the fetching advertisements of "silverplating outfits," with pretty much everything, including a princely revenue, "free," he invests all his spare dollars and runs in debt, without even posting himself on the A B C of the business, and finally throws it over, sells the outfit for junk, turns his tanks into chicken coops and retires from business, a sadder and wiser man. The fact is, silverplating, and all electroplating, is like fishing as described by Isaac Walton; "an art to be learned by practice or long observation, or both." The silverplater should begin in a very small way and advance by degrees, studying every step carefully as he goes, and mastering the plater's "art and mystery" with as little loss by error and bad work as possible.

We will suppose that the goods have been cleansed and polished and made ready for the plating tank, as previously described, including the preliminary copper coating, and are now ready for the silvering process. Careful platers, before putting the goods into the silvering tank, transfer them from the coppering solution to an alkaline mercury solution for a few minutes, to give them a film of mercury, which greatly increases the attraction of the goods to the metal in the silver solution and secures a good wearing coat. This is called "quickening."

ALKALINE QUICKENING SOLUTION

In a stoneware vessel, slowly dissolve mercury in dilute nitric acid, add enough strong solution of potassium cyanide to precipitate the mercury in the form of a black mud, and then add cyanide slowly, stirring with a stick, to dissolve the precipitate. When the liquid is clear, dilute with distilled water to make 1 gal. Keep in a stoneware jar, securely covered. Be sure to remember that cyanide is a deadly poison and cannot be too carefully used. It should not be touched with the hand or the fumes inhaled.

Accurately weigh the goods in the platers' scales, taking care not to touch the former with the hands, and place them immediately in the silverplating tank, using hooks of silver wire. The tank has been made ready with the proper amount of solution, and the anodes adjusted with due regard to the proper amount of surface to be treated, remembering that the anode surface should be slightly in excess of the surface of the cathodes. The anodes should be of pure annealed silver, not standard or coin silver.

English and German platers prefer to make their own solutions from the metal, but it is better, especially for the beginner, to take as few chances with the manipulating of chemicals as possible, and to buy the prepared salts (that is to say, combinations of acids and metals, as nitrate of copper, cyanide of silver, etc.) and make the solutions therefrom. The surest way for the beginner is to obtain his solutions from the manufacturer. It may cost a little more, but it saves time, work and money in the end to put as much of the burden as possible on the big factories.

SILVER SOLUTION

Dissolve 1 oz. of chloride of silver (always use the highest grade of chemicals) in 1 pt. of distilled water, or larger amounts in the same proportion. In a separte vessel, dissolve 1 oz. of potassium cyanide, 99 per cent pure, in ½ pt. of distilled water. Add the cyanide gradually and slowly to the chloride, stirring carefully with a hardwood stick or a glass rod. Adding the cyanide too rapidly prevents a complete union with the silver, some of which is lost. It is better to pour in a spoonful or two of the cyanide at a time and stir as long as a precipitate is formed, then add more cyanide and stir, and so on. If the precipitate has settled down in a mass, or curd, leaving the liquid clear, the chemical action has been complete and the precipitate, cyanide of silver, contains all the silver that was contained in the chloride. If the liquid, however, is cloudy, add a little more cyanide solution drop by drop until the cloudiness disappears. If the liquid has a brown tint it shows an excess of cyanide, which is corrected by adding a few drops of silver chloride until the liquid clears. When the cyanide curd has settled, pour off the liquid, which, if at all cloudy, may be saved in a crock for the recovery of the silver. Wash the curd with clean water several times and dissolve it in a solution of potassium cyanide, using about one-fifth more than actually required to take up the precipitate. Add distilled water in the proportion of 1 gal. to from 1 to 5 oz. of silver, as required. Filter through muslin. This is the cyanide solution of silver used in the plating bath. Work at a temperature of about 60° F. Do not expose the solution to strong sunlight.

STRENGTH OF SOLUTION

An ounce of silver chloride contains about 0.753 of an ounce of pure silver, and 132.8 oz. of chloride contain approximately 100 oz. of silver. Knowing the quantity of chloride used, and the number of gallons of solution, the strength of the latter is easily calculated. The solution may vary from 2 to 6 oz. of silver to the gallon, but should not fall below the former figure. A weak solution will work slower than a strong solution and will require a higher voltage. The amount above provided for is merely experimental and may be used for small articles, like a watch case, in a stoneware vessel, using a single Smee or Daniell cell.

A battery of 4 or 6 cells, gallon size, would make a good practical working outfit for silverplating, and could be so connected as to give an e. m. f. of 4 to 6 volts, which would be sufficiently strong for a solution such as above described, with sufficient anode surface. It should be remembered that the stronger the solution and the larger the anode surface, the less resistance is offered and the less voltage is required. The current density should be about 1 ampere to 60 sq. in. of surface to be coated.

The anodes should throw off silver at a rate proportionate to the silver deposited. In other words they should feed the solution and prevent it from becoming impoverished. This they are only able to do while there is free cyanide in the solution. Exposure to sunlight and air causes the cyanide to form a union with the carbon dioxide in the air, with a resultant loss of free cyanide in the solution. This is occasionally supplied by adding a few drops of cyanide solution and stirring the solution in the tank. No other chemicals whatever should be added to the solution.

The anode rod should be moved about in the solution occasionally, to aid the process of electrolysis. In large shops, the anode rods are attached to a jig frame which is kept in motion. In the same manner the cathode rods should be occasionally jarred to prevent the formation of bubbles or scum.

A lack of sufficient free cyanide is indicated by a coating of black slime on the anodes, though this may be caused also by dust and dirt in the solution if it is allowed to stand many days before being used up. The cause may be detected by a careful examination of the anodes. An excess of cyanide causes the silver to be deposited too rapidly, as shown by the rough surface and the ragged edges of the anodes.

The plating tanks should be covered when not in use, and never exposed to strong sunlight. If the shop is located in the neighborhood of factories, coal yards or railroads (a location which should be avoided if possible), there will be more or less dust and soot, and the tanks will get foul if the solution is not frequently renewed and the tanks cleaned.

The time to be allowed goods in the plating tank depends upon the strength of the solution, the current and the thickness of the coat desired. If the external resistance (of the wires, anodes and solutions) is high, or if the anodes are in bad condition, the wires small and the solution weak, the amount of silver deposited will be lessened by these causes.

The anodes, it should be remembered, feed the solution in proportion as the metal it contains is deposited on the cathodes. If the anodes, therefore, are of sufficient surface, and in good condition, that is, kept from accumulating a slime which prevents them from throwing off their metal, the solution should last indefinitely. It should, however, be occasionally taken out and filtered.

An old rule in silverplating is to allow a current of 2 amperes to each 100 sq. in. of surface to be plated, and then remove the goods and weigh to ascertain the quantity of silver deposited in an hour. By watching the deposit carefully the plater will soon become familiar with the changes that occur under varying conditions. In two or three minutes after placing them in the bath the goods should take on the whiteness of the anode plates and become beautifully frosted with grains of silver. If the deposit turns to gray and bubbles of gas arise from the slinging wire, it is an indication that the deposit is going on too fast, which may be corrected by increasing the resistance or reducing the current. If the deposit assumes a bluish tinge, it is an indication of weakness of current or of solution, in which case the anodes may be moved up closer to the goods, or, if necessary, a higher voltage may be used.

Beware of doctoring the solutions with brightening compound or any other chemical than those named in the directions given on another page for preparing the solutions. An occasional visit to other plating shops would be likely to greatly increase the beginner's stock of knowledge, and perhaps save him some costly experience.

When the goods have remained a sufficient length of time in the plating tank and the examination shows a frosty coating of dead white silver they should be taken out, rinsed, dried in hot boxwood sawdust and polished. Great care should be used in every stage of the work. The rinsing should be done in clear soft water, used hot, and the drying in clean boxwood sawdust, as this is free from injurious acids or resins, which, if present, would stain the silver. See, too, that the sawdust is dry but not charred. The article to be polished is then washed in warm mild soap-suds, rinsed in running water and dried in hot air. The article must be handled carefully, as the matt or frosted surface is easily tarnished.

The appearance of the newly plated goods before polishing is that of a frosted or cream-white surface. Should they have a yellowish tinge when taken from the vat, plunge them immediately in a warm dilute solution of potassium cyanide, rinse and dry as above directed. Silver goods should not be put on the brass wire scratchbrush, as it makes them brassy. Use a wood-fiber brush and then a soft muslin rag with rouge composition. Give the goods an even, equal pressure on all parts, without scrubbing hard on one spot. Move the article about from side to side on the rag and do not let it get heated. When a good bright polish has been put on, wash out the rouge, using hot, soapy water and a soft wool rag. Rinse in hot, clear water and dry, and then mop again with a soft swansdown muslin rag. Never put rags or mops down on a bench where they will gather dust or grit, but keep them in tightly covered boxes.

If the silver coating blisters under the polishing, there is nothing to do but strip it as directed elsewhere, and replate.

A number of small articles, such as buttons, buckles and the like, may be plated without being separately slung on wires, by placing them in an aluminum basket and shaking them about during the process. They may be dipped in the stripping acid, rinsed and dried in the same manner. It is a difficult process to connect the basket with the cathode rod and to keep the contents shaken up in an ordinary plating tank, but it may be done by an expert. A special plating machine, something like a barrel churn, is made for this sort of work and has been described before.

REPLATING STEEL TABLE KNIVES

The following simple and practical method of replating old steel table knives is given by an expert:

The knives are first examined as to condition and badly worn or pitted ones are thrown out. The rest are cleaned in a hot solution of soda or potash. The silver is then stripped in a solution of 1 lb. of potassium cyanide and 1 lb. of caustic soda in 1 gal. of water, in a plating tank. The knives are hung in a bunch as an anode by an iron wire, and a piece of cleaned sheet steel is used for the cathode, this operation being the exact reverse of plating. A strong current up to 6 volts may be used, as the cyanide does not attack the steel surface. The silver is deposited on the sheet steel in the form of a spongy coating, some of which falls to the bottom of the tank. The steel may be taken out occasionally and brushed, and so get a better deposit, the silver failing to adhere when the coating is heavy. Shake the knives occasionally to facilitate throwing off the silver. By filtering the solution, all the silver may be recovered, which is then melted in a crucible. If it is found that there is a nickel plate under the silver, which will resist the stripping process, the grinding wheel must be resorted to for its removal.

If on examination the goods appear to be pitted, which might not show before stripping, they are "cut down" on a wood wheel coated with glue and carborundum or No. 100 emery, followed by a leather-covered wheel and No. 150

emery or finer carborundum, and finished with felt and flour emery. The use of emery paste is discouraged, as it gets a film of mineral grease on or into the goods that is very hard to remove.

The knives are now dropped in a solution of 8 oz. of sal-soda in 1 gal. of water, to prevent them from rusting. After dipping for a few minutes in the hot lye (if scoured with paste, gasoline is necessary) the articles are dipped in cold water and scoured with pumice and a stiff bristle brush. Use dilute sal-soda with the pumice to keep the steel from rusting. This is much better and safer than the poisonous cyanide. All stains and rust spots must be carefully scoured out, emery flour being used if necessary. After thorough scouring, rinse in cold water, and put through the potash solution to remove grease. The knives should not be left out in the air, or they will rust, but if the plating tank is not ready they should be placed in the sal-soda solution. Do not touch with the hands. After plunging them in the lye, the knives are rinsed in cold water and passed quickly through a dip composed of 1 pt. of muriatic acid to 1 gal. of water, at a temperature of about 70° F. The goods should now be free from all grease and oxidation, or, in other words, chemically clean. This can be readily determined by the way water acts on the surface; if it wets evenly it is clean, but if it shows any repulsion or "duck's back" it must go into the potash again.

To get the best results in plating, use two baths or "strikes" before the third or final plate. The first solution is made as follows: Potassium cyanide, 2 lb.; silver chloride, 1¼ dwt; in enough water to give a density of 20° to 24° Baumé. The silver chloride may be made by dissolving 1 oz. of fine silver in nitric acid and precipitating with muriatic acid or common salt and then washing. Instead of a silver anode, a large copper anode is preferred, as it adds enough copper to the solution to make a strong deposit while giving no coppery tint. The pressure used should not be above 2 volts. Move the knives about as they hang on the rods, to get a good deposit. In five or six minutes they should have a bright straw-colored deposit, and may be taken out and brushed.

The second bath is the regulation silver "strike" of 8 oz. of potassium cyanide and ½ oz. of silver chloride in enough water to give a density of 20° Baumé. Use a pure silver anode. Four or five minutes in this bath should be sufficient.

The goods are now ready for the final plating. For this a solution is made of 6 oz. of potassium cyanide and 3 oz. of silver chloride to 1 gal. of water. Dissolve 2½ oz. of fine silver in dilute nitric acid and convert into chloride in the usual way. Then dissolve in enough strong cyanide solution to take it up, and add water enough to make 1 gal., and dissolve in it 1½ oz. of cyanide to give an excess of free cyanide. This solution will show 7° or 8° Baumé.

The knives are now placed in the tank with silver anodes on both sides to secure an even deposit. The anode surface should exceed that of the goods. Use a weak current, not over 1 volt, and move the goods in the bath often while the plating is going on. To secure what is called a "12-dwt. deposit," or 12 pennyweights of silver to the dozen knives, they should remain in the bath from 1½ to 2 hours. When the operation has been completed, the knives are taken out, scratchbrushed and buffed with a soft rag wheel and silver rouge, care being taken not to use too much rouge. Burnishing gives an extra finish, but it takes too much time for ordinary jobs.

CHAPTER VIII

GOLDPLATING

I T WILL NOT be necessary here to describe the process of goldplating any further than relates to the work of a small shop or jeweler. Of course the large outfits are entirely unsuitable for this work. Rings and other articles of jewelry, watch cases and the like, may be plated with a plant that occupies no more space than a kitchen table. Such an outfit is illustrated in Fig. 52. The electroplating supply houses offer an outfit about as follows: One enameled iron vessel; 1 large Bunsen cell; 1 gal. of 24 or 14-karat gold solution; gold anode; connecting wire;

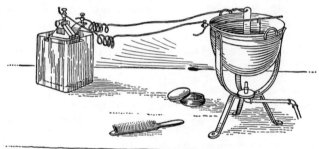

Fig. 52. Small Goldplating Outfit

scratchbrush; hard and soft gold rouge. Polishing mops one can make for himself. As this sort of work is often done in a dwelling or ill-ventilated room, it is desirable to use a battery which does not give off noxious fumes.

For goldplating, the goods must be thoroughly cleaned and prepared very much as in the processes for other metals. It is advisable to copperplate articles of soft metal,

74

as in silverplating, but quickening with mercury is unnec-
essary. Copper, silver, brass and various alloys used in
making cheap jewelry usually take a gold deposit with no
difficulty. Aluminum articles should be coppered in a solution of copper sulphate. The better the polish that is given
the goods before gilding, the better the work will result.
If a matt or frosted surface is desired, roughen the surface
with acid or a frosting brush before gilding. Some jewelers and platers make their own gold solution by dissolving
pure gold in a heated solution of potassium cyanide, and
passing an electric current through the solution until sufficient metal is dissolved. The process will not be given here
in detail, as in practice it would be found wasteful by any
but an expert. Solutions of known strength may be bought
of the supply houses at very little above the value of the
gold and these save time and labor and are always reliable.
As in plating silver, it is not desirable to put several articles
of different metals or alloys in the plating bath at once.
Several copper articles, or several silver articles, may be
worked at the same time.

In a small earthen dish or crock 6 or 8 in. in diameter,
for the plating bath, the articles are suspended by fine copper wire from a cathode rod of copper the size of ordinary
fence wire. See that the connections to the articles are
good, and that anything of the link or chain character is
attached to the wire in several places. The anode should be
of pure gold plate not less than 0.06 inch in thickness. Leaf,
or any ribbon thinner than this, is apt to be worn away and
wasted. A platinum wire should be used for the anode.
As in silver, the anode surface should be slightly in excess
of the surface of the cathode. It should be frequently
moved in the bath, and always taken out and dried when
not in use.

An excess of current, too large an anode surface, or a
deficiency of free cyanide in the solution, is indicated by a
gathering of slime on the anode. If the plater knows that

the solution is not overworked or exhausted, he should look carefully to the current and the relation of anode and cathode.

The temperature of the gilding bath should be from 140° to 180° F. when working. At a lower temperature the deposit is lighter colored, and below 120° it will be of a straw color or brassy tint. As the internal resistance of the bath is reduced by heat, the color darkens, and above 160° it takes on a coppery hue. A fresh solution deposits gold of a very light tint, while as it grows older, the tint grows darker. The alloys are acted upon by the solution, as are the copper wires, and they gradually impart an alloy to the coat deposited.

For small work, a single cell, working at ½ volt and a fraction of an ampere, will give sufficient current. As in silverplating, when the current is too strong, the work "burns" and the gold is deposited in a loose, powdery form, reddish brown in color. It is best to have a rheostat, in gilding, even with a small outfit, to insure proper voltage, for this is one of the places where guesswork does not pay. If a job is "burned" so as to give it a bronze tinge, it can only be made right by stripping and regilding. Sometimes, however, a dark brown deposit of gold is caused by excess of free cyanide and a deficiency of gold in the solution, a condition which is remedied by adding cyanide of gold to the solution.

Another condition affecting the color of the gilding is the metal which forms the base, which may be said to shine through the extremely thin film of gold and impart its own hue to the gilding; hence, if other conditions are right, the gilding, which may look at first very dark on copper, or light on silver, will be apt to improve as more gold is deposited. A knowledge of the behavior of gold in depositing on other metals will permit the expert plater to produce some very attractive effects. Thus, on freshly deposited copper and silver, on frosted surfaces, and where frosted and burnished effects are combined, a thin gilding often produces beautiful results, which may be given fanciful names such as gold of Ophir, Ormus gold, Alaska gold, etc. There are several popular fallacies in connection with the color of gold from various localities and as regards certain so-called "golds." Thus, it is commonly believed that African gold is of a coppery hue, while California gold is light in tint; that "rolled gold" is anything but gilt, and usually a very low grade of gilt at that. But these are simply fallacies, though so dear are they to the popular heart that it is useless to attempt to destroy them.

If the plater has progressed far enough to handle several plating vessels with different solutions, he may work with various alloys, or copper, silver, etc., which will give almost as many hues to the goods (only portions of which may be immersed in the bath) as could be given by using pigments applied with a brush. For example, a green gilt is produced by adding a copper cyanide solution, or the use of a copper anode until the desired tint is obtained. Rose pink is effected by scratchbrushing the first filmy deposit of gold, then giving it a thin coating of copper, and finishing with a blush of gold on the face of the copper. These, however, may be called eccentricities for the amusement and diversion of a wealthy alchemist, rather than the work of a commercial electroplater.

Heavy gilding is done, not by allowing the article to remain long in the bath, but by taking it out often and scratchbrushing it to remove the loosely adhering particles of gold and then returning it to the bath for another coat.

Small articles of steel, if they have a highly polished surface, may be thoroughly cleansed and put through the gilding bath suspended in a platinum basket, as described for buttons, etc., in silverplating. When this process is followed, the gold on the platinum basket may be recovered by hanging it on the anode rod in the bath.

A metal egg cup or salt cellar or other cup-shaped vessel

may be gilded on the inside by filling it with the solution and connecting it with the cathode rod, while a small plate of gold forming the anode, is suspended in the solution.

Gold is deposited more rapidly than nickel or copper but not as rapidly as silver, the ratio being 37.8 gr. per hour with a current of one ampere. When it is calculated that sufficient gold has been deposited, which, after all, can only be judged by experience, the articles are taken out, rinsed in clear water, scratchbrushed, dried by rubbing in hot boxwood sawdust, and weighed. Goldsmiths' balances showing tenths of grains, will be required and should be kept in a separate closet assigned to the goldplating outfit. The articles having been carefully weighed before plating and after the process is completed, the difference shows the amount of metal deposited.

In actual practice the work will be something like this: A charge of say $1.00 is made for gilding or triple-plating a watch fob, badge or some similar article. (The term "triple plate" is like rolled gold, it does not mean much of anything, but it sounds well and people like it). This will allow for 5 gr. of gold, and $.79 for work, there being 23.22 gr. of pure gold in one dollar. A piece of gold plate having a larger surface than the fob or badge, supposing it to be the only article to be gilded, is suspended on the anode rod and the fob, prepared and weighed as already described, is wired on the cathode rod. If several pieces are gilded at the same time, of course the expense of the process in time is proportionately reduced. After remaining in the bath 10 minutes, the fob is taken out, scratchbrushed and weighed. If it is found to have gained, say, 2 gr. in the 10 minutes time, it is clear that in 15 minutes more it should take on the required additional 3 gr., and it is put back for that time.

In gilding bracelets, chains or other similar articles of jewelry, care should be taken to see that all joints are well soldered, so that they do not become filled with the plating solution and cause errors in the calculation of the gold deposit as indicated by the weight.

Jewels in rings, society emblems, etc., should be removed before gilding and placed in a small box properly marked and kept under lock and key. They may be glass but that, of course, is a professional secret. All soft metal

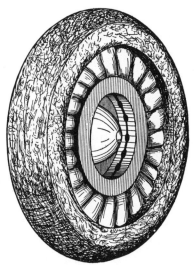

Fig. 53. Lathe Buff for Goldplating

and alloy goods and articles which have been repaired should be copperplated before gilding.

Waste from the gold table should be saved for the recovery of the gold it contains. This is done by letting the water settle and carefully dipping off the clear water. Evaporate the rest (if a solution, all should be evaporated) in a smooth porcelain-lined kettle or dish. To the remaining salt or residuum add an equal bulk of litharge (lead monoxide) and mix well. Calcine in a crucible and keep at a glowing heat over a gas burner until all the litharge has been decomposed, leaving a button of lead and gold with the silver, copper and other metals of the alloy. This may be sent to the assayer, who will allow for the precious

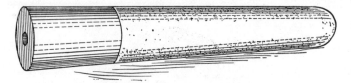

Fig. 54. Lathe Buff for Goldplating

metals it contains. Unless the plater is an expert chemist, however, he had better not undertake the delicate process of reduction.

The tools and articles used in goldplating should be kept for that purpose only, and not mixed up with others. The brushing and cleaning may be done entirely by hand, but if much work of this character is done a small lathe will be found useful. A small scratchbrush of fine brass wire,

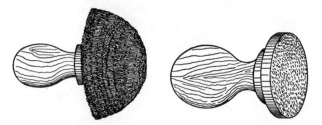

Fig. 55. Hand Buff   Fig. 56. Hand Buff

one of crimped wire, several small brushes, mops of swansdown muslin, soft felt, chamois skin, etc., for the lathe and hand work, with a supply of French rouge compound, are among the necessaries. Several kinds of buffs are shown in

Figs. 53, 54, 55 and 56. Small brushes or mops on pencils and penholders which will answer all the purposes of more expensive articles, may be made up by the gilder. Burnishers are made of steel and agate, and economical jewelers in the country are said to use their wives' manicure sets for this purpose. All repairs, filling of pin holes, removal of dirt, etc., must be done before gilding.

Gilded work should have only sufficient scratchbrushing to take off the loose metal. The operator should also hammer down the minute particles into a smooth, even coat. After this is done, scour with soft felt, swansdown muslin and chamois, using a little fine rouge composition. In using rouge, one must take care to wash it all out of articles that have engraved work or crevices which are not easily reached with the brushes. Chains, rings, thimbles, etc., are best polished by using small bobs of fine felt or chamois on the polishing lathe.

In the process of brushing, a lubricant should be used both to prevent the gold wearing away and to keep the brass from depositing on the gold. English platers always recommend stale beer, but a weak solution of linseed is really better.

If a lathe is used, the brush should be protected by a hood to prevent the waste of gold in splashing. A few minutes may be profitably spent in watching an expert polisher at work. Notice the brushes and bobs used for each class of work, and how the article is held against the brush so as to get the smooth brushing motion and not the direct blow of the bristles, and how spattering is avoided as much as possible simply by the position in which the article is held.

BURNISHING GOLD

Burnishing is an art in itself. Much harm can be done by using a rusty steel burnisher, or one with a sharp edge. Rust is corrosion of the metal by the oxygen in the air, or from acids, and the result is a roughening of the bright surface. Steel burnishers should have all the care one gives to a good razor and perhaps more, for a rough burnisher may spoil a job. The common shapes of steel burnishers are show in Fig. 57. Steel burnishers should have their bright surfaces kept in good condition by frequent rubbings on a soft leather pad with putty powder. The

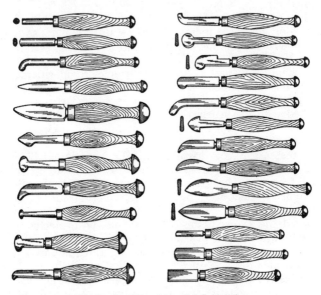

Fig. 57. Steel Burnishers for Goldplating

burnishers should never be put away without being wiped with an oiled cloth, in which they should be wrapped to protect them from acid fumes and moisture. Burnishers of agate or bloodstone cost more than those of steel, but they do not rust. They also have a smoother and harder surface than the best steel, and so give a brighter polish to the goods. Especial care must be given to agate burnish-

ers to see that the corners do not get chipped, which gives
them a cutting edge that is destructive to the articles on
which they are worked. They should always be wrapped
in muslin or linen when put away, and carefully examined
under a glass before used. These tools cost from $1.00 to
$1.25 for steel and $1.50 to $2.50 for agate, etc. Lend your
plating dish, if you must, for a preserving kettle, and let
thieves run off with your gold solution, but never, under
any circumstances, lend your burnishers.

The following directions will help the beginner: Regard
the article on which you are working as a soft material, as
indeed it is, covered by a very thin film of an even softer
material. The burnishing tool simply presses down these
particles, mats them together and so smooths and brightens
the surface, much as a boy makes a hard snowball out of
flaky snow, or the glazier spreads and packs putty against
the edge of a sheet of glass. Remember that the object
from the beginning of the polishing is not to wear away the
gold, but to smooth it down. With this purpose in view the
ring or other article is held on a soft pad with one hand,
while with the other the burnisher is pressed hard on the
surface with swathing strokes overlapping each other and
always in the same direction. To avoid friction, the surface may be lubricated with weak castile-soap suds freshly
made, using the steel burnisher first and finishing with
agate. When a sufficiently bright and uniform surface is
obtained, rinse in hot water and dry with a clean soft linen
cloth.

For polishing steel and agate burnishers, a fine Lake
Superior stone is used (one of clear flint, which should be
kept exclusively for this purpose, and is to be used only for
removing rust or sharp edges), and a buff, which is like a
hone made of soft leather. To make this, get a piece of soft
dressed leather (calf skin will do, but deer skin is better;
chamois is too thin), say, 2 by 4 in. Boil it in soft water to
remove chemicals, dry it quickly and soften by rubbing.

Then glue it to a block of beech or maple 1 in. thick, 5 in. long by 3 in. broad. The dimensions are not important, but are given as a general guide. Use warm glue, but thin enough to strike through the leather, and press down with a clean block and weights. Keep it covered. If the buff and stone are permitted to gather dust they become abrasives rather than polishers, in fact, grindstones. Therefore they should be kept in boxes or wrapped in a bit of muslin. The value of tools depends very much on the way they are kept.

Burnishing requires a good light, a careful eye and a steady hand. It takes skill, which is a little word that includes everything. As to the brushing, there are many details of the work that cannot be given in the printed directions and which can only be learned by careful practice and the observation of the work of an expert.

Burnishing should be done in a room that is free from dust. The windows should be protected by gauze screens. Do not try to burnish by artificial light, and it is better to have no gas or stove in the room, or furnace heat, as these all give out gases injurious to fine polished surfaces, especially silver.

CHAPTER IX

THE NICKELING of molded work is a more delicate process than the plating of metal goods, but not different after the process has been learned. A wax mold of the article having been prepared, it is dusted with black lead and placed in the copper bath for a thin film of copper. The mold must be free from undercutting, or interstices which would prevent a free separation of the mold and the matrix. The copper matrix is then removed from the mold and placed in the nickel vat and a coat of nickel of the desired thickness applied.

TO MAKE COPIES OF WAX MOLDS

Fine copies of wax impressions can be made in a very simple manner as follows: If the article is no larger than an ordinary medal or brooch intaglio, take an ordinary tumbler and partially fill it with a strong solution of sulphate of copper. Then make a porous cell by rolling a piece of stiff brown paper around a piece of broom handle and fasten the edge with sealing wax, putting in a bottom in the same way. A Welsbach burner box will serve very well. Make a solution of 1 part of oil of vitriol and 5 parts of water (adding the vitriol slowly to the water), and pour this mixture into the porous cell. Wind the end of a small copper wire around the end of a bit of zinc the size of a pencil and drop the zinc into the porous cell. Attach the other end of the wire to the wax impression, which is prepared thus: Pour melted beeswax on the medal, and remove carefully when cold. Then dust the mold with black lead and polish with a stiff camel's-hair brush. Suspend the porous cup and the wax mold in the tumbler, connect with a battery (one cell) and run about

12 hours. Take out and polish, and a good copy should result.

OXIDIZING COPPER, BRONZE AND SILVER

The process of oxidizing is simply the corroding or rusting of the bright metal surface to give it a frosted appearance. It may be done in various ways to give the effect desired. A simple method is as follows: Dissolve ½ oz. of sulphate of potash in 1 gal. of water, and use as a dip solution at 130° F. in a vitrified earthenware crock. The high light is then brought out by slightly touching the work to a felt wheel charged with rouge. Sewed buffs are also used with good effect, and for a rougher finish the steel or brass wire brush is used.

BRONZING

Real bronze is an alloy of copper and tin, with a small addition of zinc, and sometimes of aluminum. A "bronze effect" is often given to iron and copper goods by the use of an alkali dip. Bronzing solutions in the electroplating tank may be made to give a passable bronze luster by the addition of copper with expert handling. The following alkaline coppering solution is often called a bronze solution and works in a very satisfactory manner.

BRONZE COPPERING SOLUTION

Dissolve 2 oz. of copper sulphate in 1 qt. of hot distilled water, and add it to ½ gal. of water in which has been dissolved 4 oz. of potassium carbonate; add 2 oz. of liquid ammonia and stir until the green precipitate is entirely dissolved. Make a solution of 6 oz. of potassium cyanide in ½ gal. of distilled water; mix with the above and filter through felt or double muslin. Work at 80° to 100° F. To deepen the bronze tint dip the coppered goods in a solution of sal ammoniac.

SILVERPLATING WITHOUT A BATTERY

A simple and practical home silvering process is as follows: Dissolve two silver dollars or their equivalent in 2 oz. of nitric acid and add slowly 4 oz. of common salt dissolved in just enough water to take up the salt. Decant the liquid and wash the silver-chloride precipitate with water. Dissolve it in a solution of 2 oz. of potassium cyanide, and 3 oz. of hyposulphite of sodium in 6 oz. of water. Filter the solution and make up with distilled water to 2 qt. Clean the articles to be plated with lye and hang them by a stick of lead in a stoneware crock and boil 10 or 20 minutes in this solution. Be careful not to breath the fumes of this solution, which is a deadly poison.

CLEANING BRASS AND STEEL

For exhibition purposes, but not for plating, brass and steel articles may be quickly and cheaply cleaned by rubbing with vinegar and salt, or dilute oxalic acid. Wash immediately and polish with tripoli and sweet oil.

TO PREVENT TARNISH ON SILVER

Brush alcohol in which a little collodion has been dissolved over silverware that is to be set away for any length of time. The thin invisible coating left by the solution can be readily removed by dipping the article in hot water.

METALLIZED FLOWERS

The making of the so-called "metallized roses" and other natural flowers is upon the same principle as plating molded work. The process is patented and the secret is carefully guarded. The general method of the work, however, is understood to be as follows: The rose or other bud is dipped in a mucilage which stiffens the petals, and is then dusted with plumbago. It is then given a copper coating in a cold bath and the copper is then worked in

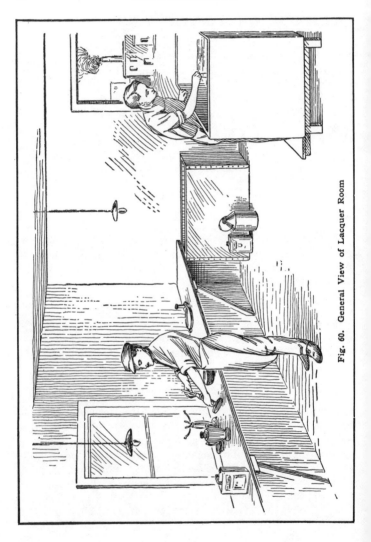

Fig. 60. General View of Lacquer Room

Fig. 58. Dip Lacquering

Fig. 59. Brush Lacquering

various tints, as may be required, and colored by hand. A thin gilding may be given if desired. If the plating jeweler is successful in his experiments, he may strike a very profitable business.

LACQUERING

Many plated articles, such as buckles, trimmings, etc., for the trade, are lacquered to protect the plating and give a higher and more lustrous finish. The lacquer should be thin, colorless and of good quality. It should be applied carefully to insure a good job. Lacquers are of two kinds, dip lacquer and brush lacquer, and the names will indicate their use.

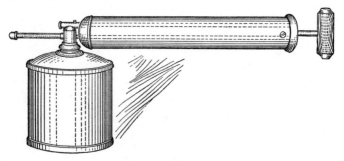

Fig. 61   Hand Sprayer

The dip lacquer should be kept in a glass or stoneware tank or jar, with a tight cover. If much work is done a tin-lined wooden tank is used. The goods should be cleaned as for plating. Hang them on a wire so that they do not touch each other, dip and let them drip over the tank The method is shown in Fig. 58. The lacquer should run well, and not be thin enough to be iridescent (rainbow-colored) nor thick enough to show the drip—that is, the accumulated hardened drops at the bottom. Work at about 100° F. Dry in air free from dust at 100° F. to 140° F.

If the lacquer has to be thinned, use a thinner of the same grade.

Brush lacquers are used thicker than dip lacquers, but as this is possible without showing iridescence give a flowing coat with a soft lacquer brush as illustrated in Fig. 59. See that no rouge, or any composition of any sort or any greasy deposit is left on the goods, as this will give a milky appearance. Do not touch the articles with the hands.

A general view of the lacquer room is shown in Fig. 60.

Manufacturers of metal goods are now successfully applying lacquer by the use of sprayers operated by compressed air. A hand sprayer, Fig. 61, is also made which answers very well where compressed air is not available. The great advantages of the sprayer where much lacquering is done lies in the fact that a much smaller quantity and a thinner body of lacquer is required than by the dip or brush methods.

REMOVAL OF STAINS

To remove stains of copper sulphate, or salts of mercury, gold, silver, etc., from the hands, wash them with a very dilute solution of ammonia, and then with plenty of water; if the stains are old ones, they should be rubbed with the strongest acetic acid, and then treated as above.

Grease, oil, tar, etc., may be removed from the hands or clothes by rubbing with a rag saturated with benzine, turpentine, or carbon bisulphide.

CHAPTER X

FIRST AID TO THE INJURED

THE CHEMICALS with which the electroplater has to do, are, many of them, deadly poisons, and their combinations often produce corrosive and poisonous gases. Familiarity breeds contempt, as we know, but it is criminal carelessness, and nothing less, if the plater allows his familiar use of deadly poisons to endanger the lives of himself and his workers.

Have the following list of antidotes and the directions for their use typewritten and tacked up on the door of a "First Aid" cabinet on the wall. Study it and memorize the leading points, because when an accident happens with the hydrochloric acid, the sulphuric acid or the like, there is no time for study, and if the right thing is not done on the instant, it is a case for the coroner.

ANTIDOTES FOR POISONS

In case of poisoning, send or telephone for a doctor, and take remedial measures as quickly as possible, as follows:

Nitric, Hydrochloric or Sulphuric Acids.—Give as much tepid water as possible to act as an emetic, or milk, the whites of eggs, some calcined magnesia, or a mixture of chalk and water. If those acids, in a concentrated state, have been spilled on the skin, apply a mixture of whiting and olive oil. If the quantity is very small, simple washing with plenty of cold water will suffice. A useful mixture, in cases of burning with strong sulphuric acid, is formed with 1 oz. of quicklime slaked with ¼ oz. of water, and then adding it to 1 qt. of water. After allowing to stand some time, pour off the clear liquid and mix it with olive oil to form a thin paste.

Potassium Cyanide, Hydrocyanic Acid, etc.—If cyanides, such as a drop of an ordinary plating solution has been accidentally swallowed, water, as cold as possible, should be run on the head and spine of the sufferer, and a dilute solution of iron acetate, citrate or tartrate given. If hydrocyanic-acid vapors have been inhaled, cold water should be applied as above, and the patient be caused to inhale air containing a little chlorine gas, which can be kept in solution in water in a tightly stoppered bottle. Chlorine solution may be purchased in bottles. Never dip the arms into a plating solution to recover any work that has fallen off the wires, because the skin often absorbs cyanide liquids, causing painful sores. In such a case, wash well with water, and apply the olive oil and lime-water mixture.

Alkalies.—These bodies are the opposite of acids in character, so that acids may be used as antidotes. It is preferable to employ weak acids, such as vinegar or lemonade; but if these are not at hand, then use exceedingly dilute sulphuric acid, or even nitric acid diluted, so that it just possesses a decidedly sour taste. After about ten minutes take a few teaspoonfuls of olive oil.

Mercury Salts.—The white of an egg is the best antidote in this case. Sulphur and sulphureted hydrogen are also serviceable for the purpose.

Copper Salts.—The stomach should be quickly emptied by means of an emetic, or in want of this, the patient should thrust his finger in the back of his throat, so as to induce vomiting. After vomiting, drink milk, white of an egg, or gum water.

Lead Salts.—Proceed as in case of copper salts. Lemonade, soda water and sodium carbonate are also serviceable.

Acid Vapors.—Admit immediately an abundance of fresh air, and inhale the vapor of ammonia, or a few drops of ammonia may be put into a glass of water and the solution drunk. Take plenty of hot drinks and excite warmth by friction. Employ hot foot baths to remove the blood from the lungs. Keep the throat moist by sipping milk.

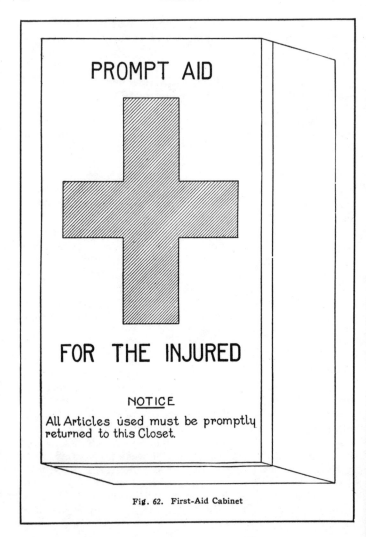

Fig. 62.   First-Aid Cabinet

FIRST-AID CABINET

In a handy little closet on the wall (you can make one out of a soap box) as shown in Fig. 62, keep the supplies listed below, omitting none of them because you may possibly have them somewhere about the shop. The whole outfit will not cost $2.00 and it may save a life.

8-oz. bottle of olive oil.   Label—For acids and alkalies.

4-oz. bottle of calcined magnesia.   Label—For acids.

4-oz. bottle of lime and oil dressing.   Label—For burns —Shake well.

4-oz. bottle of ammonia.   Label—For acid vapors.

½ oz. cream of tartar.   Label—For mercury salts.

¼ lb. vaseline.   Label—For burns.

½ lb. absorbent cotton.

4 oz. iron acetate.   Label—For cyanides.

CHAPTER XI

WHEN THE beginner has finished several samples of work, which will pass the inspection of a good judge, he is then ready to set up in business. A good scheme is to start out with some simple article experience has shown can be done easily and well, such as a collar button, a belt buckle, watch fob or perhaps a souvenir spoon if a good original design can be made. Put a fair price on the article and show it to all your friends. Advertise it in the local paper and give the editor one or more for his own use. Get some cards neatly printed with just your name, business and location and distribute them. Do not load down your card or billhead with a lot of useless printing. Work the automobile garages and try to get their business. Remember, you must make your business known or people will not come to you. Don't hide your card under a bush.

Watch your own work carefully, and that of your assistant, if you have one, still more carefully. Get a young fellow of sense and ambition and pay him by the week. Piece work is apt to be slighted and does not pay.

Keep careful accounts, and especially a book of receipts and description of goods. This prevents annoying errors. Have an understanding with a larger shop that does good work to take jobs off your hands that you are unable to undertake, and then you need not turn a job away because you have not the facilities for taking care of it. By watching how the bigger shop conducts its business you will get points for managing your own. If you make an estimate on a job that brings you a loss, you may tell your customer, but do not attempt to raise the price. It will please him to know that he gets more than he pays for, and you can charge the difference up to experience.

Alphabetical Index

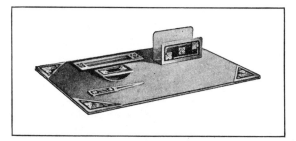